Essentials of Cosmology

Essentials of Cosmology

Edited by
Kent Buchanan

Larsen & Keller
www.larsen-keller.com

Essentials of Cosmology
Edited by Kent Buchanan
ISBN: 978-1-63549-681-9 (Hardback)

© 2018 Larsen & Keller

Published by Larsen and Keller Education,
5 Penn Plaza,
19th Floor,
New York, NY 10001, USA

Cataloging-in-Publication Data

Essentials of cosmology / edited by Kent Buchanan.
        p. cm.
Includes bibliographical references and index.
ISBN 978-1-63549-681-9
1. Cosmology. 2. Universe. I. Buchanan, Kent.
QB981 .E87 2018
523.1--dc23

For more information regarding Larsen and Keller Education and its products, please visit the publisher's website www.larsen-keller.com

Table of Contents

Preface

Cosmology is a vast subject. It studies the origins, history, evolution and growth of our planet. The subject uses laws of physics, chemistry, etc. to understand the complexities present in the universe. The subjects studied under this field are dark matter, celestial objects, dark energy, worm-holes, etc. The different branches of cosmology are mythological cosmology, physical cosmology and philosophical cosmology. This book presents the complex subject of cosmology in the most comprehensible and easy to understand language. The topics covered in it deal with the core subjects of the field. The textbook is appropriate for those seeking detailed information in this area.

To facilitate a deeper understanding of the contents of this book a short introduction of every chapter is written below:

Chapter 1- This chapter will provide an integrated understanding of cosmology. It is the scientific study of the origin and evolution of the universe. Astronomers as well as philosophers study cosmology. Eschatology, axis mundi, geocentric model and heliocentrism are some of the topics studied in this section.

Chapter 2- Physical cosmology is the study of the structure of the universe. It is concerned with fundamental questions about its creation and evolution. In order to develop a better understanding of physical cosmology, it is important to understand big bang nucleosynthesis, absolute time and space, dark energy and Friedmann–Lemaître–Robertson–Walker metric. This chapter is an overview of the subject matter incorporating all the major aspects of physical cosmology.

Chapter 3- The observable universe consists of matter that can be observed from Earth. Particle horizon, Hubble volume, redshift survey and comoving distance are some of the significant and important topics related to cosmology. The following chapter unfolds its crucial aspects in a critical yet systematic manner.

Chapter 4- Religious cosmology is the explanation of the universe from a religious point of view. It studies the religious interpretation of the origin and history of the cosmos. The topics discussed in the chapter are of great importance to broaden the existing knowledge on cosmology.

I would like to share the credit of this book with my editorial team who worked tirelessly on this book. I owe the completion of this book to the never-ending support of my family, who supported me throughout the project.

Editor

An Introduction to Cosmology

This chapter will provide an integrated understanding of cosmology. It is the scientific study of the origin and evolution of the universe. Astronomers as well as philosophers study cosmology. Eschatology, axis mundi, geocentric model and heliocentrism are some of the topics studied in this section.

Cosmology

The Hubble eXtreme Deep Field (XDF) was completed in September 2012 and shows the farthest galaxies ever photographed. Except for the few stars in the foreground (which are bright and easily recognizable because only they have diffraction spikes), every speck of light in the photo is an individual galaxy, some of them as old as 13.2 billion years; the observable universe is estimated to have contained more than 2 trillion galaxies.

Cosmology (from the Greek "world" and "study of") is the study of the origin, evolution, and eventual fate of the universe. Physical cosmology is the scholarly and scientific study of the origin, large-scale structures and dynamics, and ultimate fate of the universe, as well as the scientific laws that govern these realities.

The term *cosmology* was first used in English in 1656 in Thomas Blount's *Glossographia*, and in 1731 taken up in Latin by German philosopher Christian Wolff, in *Cosmologia Generalis*.

Religious or mythological cosmology is a body of beliefs based on mythological, religious, History of the center of the Universe.

Physical cosmology is studied by scientists, such as astronomers and physicists, as well as philosophers, such as metaphysicians, philosophers of physics, and philosophers of space and time. Because of this shared scope with philosophy, theories in physical cosmology may include both scientific and non-scientific propositions, and may depend upon assumptions that can not be tested. Cosmology differs from astronomy in that the former is concerned with the Universe as a whole while the latter deals with individual celestial objects. Modern physical cosmology is dominated by the Big Bang theory, which attempts to bring together observational astronomy and particle physics; more specifically, a standard parameterization of the Big Bang with dark matter and dark energy, known as the Lambda-CDM model.

Theoretical astrophysicist David N. Spergel has described cosmology as a "historical science" because "when we look out in space, we look back in time" due to the finite nature of the speed of light.

Disciplines

Physics and astrophysics have played a central role in shaping the understanding of the universe through scientific observation and experiment. Physical cosmology was shaped through both mathematics and observation in an analysis of the whole universe. The universe is generally understood to have begun with the Big Bang, followed almost instantaneously by cosmic inflation; an expansion of space from which the universe is thought to have emerged 13.799 ± 0.021 billion years ago. Cosmogony studies the origin of the Universe, and cosmography maps the features of the Universe.

In Diderot's Encyclopédie, cosmology is broken down into uranology (the science of the heavens), aerology (the science of the air), geology (the science of the continents), and hydrology (the science of waters).

Metaphysical cosmology has also been described as the placing of man in the universe in relationship to all other entities. This is exemplified by Marcus Aurelius's observation that a man's place in that relationship: "He who does not know what the world is does not know where he is, and he who does not know for what purpose the world exists, does not know who he is, nor what the world is."

Physical Cosmology

Physical cosmology is the branch of physics and astrophysics that deals with the study of the physical origins and evolution of the Universe. It also includes the study of the nature of the Universe on a large scale. In its earliest form, it was what is now known as "celestial mechanics", the study of the heavens. Greek philosophers Aristarchus of Samos, Aristotle, and Ptolemy proposed different cosmological theories. The geocentric Ptolemaic system was the prevailing theory until the 16th century when Nicolaus Copernicus, and subsequently Johannes Kepler and Galileo Galilei, proposed a heliocentric system. This is one of the most famous examples of epistemological rupture in physical cosmology.

When Isaac Newton published the *Principia Mathematica* in 1687, he finally figured out how the heavens moved. Newton provided a physical mechanism for Kepler's laws and his law of universal gravitation allowed the anomalies in previous systems, caused by gravitational interaction between the planets, to be resolved. A fundamental difference between Newton's cosmology and those preceding it was the Copernican principle—that the bodies on earth obey the same physical laws as all the celestial bodies. This was a crucial philosophical advance in physical cosmology.

Evidence of gravitational waves in the infant universe may have been uncovered
by the microscopic examination of the focal plane of the BICEP2 radio telescope.

Modern scientific cosmology is usually considered to have begun in 1917 with Albert Einstein's publication of his final modification of general relativity in the paper "Cosmological Considerations of the General Theory of Relativity" (although this paper was not widely available outside of Germany until the end of World War I). General relativity prompted cosmogonists such as Willem de Sitter, Karl Schwarzschild, and Arthur Eddington to explore its astronomical ramifications, which enhanced the ability of astronomers to study very distant objects. Physicists began changing the assumption that the Universe was static and unchanging. In 1922 Alexander Friedmann introduced the idea of an expanding universe that contained moving matter.

In parallel to this dynamic approach to cosmology, one long-standing debate about the structure of the cosmos was coming to a climax. Mount Wilson astronomer Harlow Shapley championed the model of a cosmos made up of the Milky Way star system only; while Heber D. Curtis argued for the idea that spiral nebulae were star systems in their own right as island universes. This difference of ideas came to a climax with the organization of the Great Debate on 26 April 1920 at the meeting of the U.S. National Academy of Sciences in Washington, D.C. The debate was resolved when Edwin Hubble detected Cepheid Variables in the Andromeda galaxy in 1923 and 1924. Their distance established spiral nebulae well beyond the edge of the Milky Way.

Subsequent modelling of the universe explored the possibility that the cosmological constant, introduced by Einstein in his 1917 paper, may result in an expanding universe, depending on its value. Thus the Big Bang model was proposed by the Belgian priest Georges Lemaître in 1927 which was subsequently corroborated by Edwin Hubble's discovery of the red shift in 1929 and later by the discovery of the cosmic microwave background radiation by Arno Penzias and Robert Woodrow Wilson in 1964. These findings were a first step to rule out some of many alternative cosmologies.

Since around 1990, several dramatic advances in observational cosmology have transformed cosmology from a largely speculative science into a predictive science with precise agreement between theory and observation. These advances include observations of the microwave background from the COBE, WMAP and Planck satellites, large new galaxy redshift surveys including 2dfGRS and SDSS, and observations of distant supernovae and gravitational lensing. These observations matched the predictions of the cosmic inflation theory, a modified Big Bang theory, and the specific version known as the Lambda-CDM model. This has led many to refer to modern times as the "golden age of cosmology".

On 17 March 2014, astronomers at the Harvard-Smithsonian Center for Astrophysics announced the detection of gravitational waves, providing strong evidence for inflation and the Big Bang. However, on 19 June 2014, lowered confidence in confirming the cosmic inflation findings was reported.

On 1 December 2014, at the *Planck 2014* meeting in Ferrara, Italy, astronomers reported that the universe is 13.8 billion years old and is composed of 4.9% atomic matter, 26.6% dark matter and 68.5% dark energy.

Religious or Mythological Cosmology

Religious or mythological cosmology is a body of beliefs based on mythological, religious, and esoteric literature and traditions of creation and eschatology.

Philosophical Cosmology

Cosmology deals with the world as the totality of space, time and all phenomena. Historically, it has had quite a broad scope, and in many cases was founded in religion. The ancient Greeks did not draw a distinction between this use and their model for the cosmos. However, in modern use metaphysical cosmology addresses questions about the Universe which are beyond the scope of science. It is distinguished from religious cosmology in that it approaches these questions using philosophical methods like dialectics.

Cosmology (Philosophy)

Scenographia Systematis Copernicani, engraving

Philosophical cosmology, philosophy of cosmology or philosophy of cosmos is a discipline directed to the philosophical contemplation of the universe as a totality, and to its conceptual foundations. It draws on several branches of philosophy --metaphysics, epistemology, philosophy of physics, philosophy of science, philosophy of mathematics, and on the fundamental theories of physics. The term *cosmology* was used at least as early as 1730, by German philosopher Christian Wolff, in *Cosmologia Generalis*.

It can be distinguished by two types of cosmological arguments: deductive and inductive cosmological arguments. The first type has a long tradition in the history of philosophy, proposed by thinkers like Plato, Aristotle, Descartes and Leibniz, and criticized by thinkers like David Hume, Immanuel Kant and Bertrand Russell, while the latter has been formulated by philosophers like Richard Swinburne.

For Leibniz, all the plenum of the universe is entirely filled with tiny Monads, which cannot fail, have no constituent parts and have no windows through which anything could come in or go out. In his *Aesthetics*, philosopher José Vasconcelos explains his theory on the evolution of the universe and the restructuring of its cosmic substance, in the physical, biological and human orders.

Philosophical cosmology tries to respond questions such as:

- What is the provenance of the cosmos?

- What are the essential constituents of the cosmos?

- Does the cosmos have an ulterior motive?

- How does the cosmos behave?

- How can we understand the cosmos in which we find ourselves?

Eschatology

Four Horsemen of the Apocalypse by Albrecht Dürer

Eschatology is a part of theology concerned with the final events of history, or the ultimate destiny of humanity. This concept is commonly referred to as the "end of the world" or "end time".

The word arises from the Greek meaning "last" and *-logy* meaning "the study of", first used in English around 1844. The Oxford English Dictionary defines eschatology as "The part of theology concerned with death, judgment, and the final destiny of the soul and of humankind."

In the context of mysticism, the phrase refers metaphorically to the end of ordinary reality and reunion with the Divine. In many religions it is taught as an existing future event prophesied in sacred texts or folklore. More broadly, eschatology may encompass related concepts such as the Messiah or Messianic Age, the end time, and the end of days.

History is often divided into "ages" (aeons), which are time periods each with certain commonalities. One age comes to an end and a new age or world to come, where different realities are present, begins. When such transitions from one age to another are the subject of eschatological discussion, the phrase, "end of the world", is replaced by "end of the age", "end of an era", or "end of life as we know it". Much apocalyptic fiction does not deal with the "end of time" but rather with the end of a certain period of time, the end of life as it is now, and the beginning of a new period of time. It is usually a crisis that brings an end to current reality and ushers in a new way of living, thinking, or being. This crisis may take the form of the intervention of a deity in history, a war, a change in the environment, or the reaching of a new level of consciousness.

Most modern eschatology and apocalypticism, both religious and secular, involve the violent disruption or destruction of the world; whereas Christian and Jewish eschatologies view the end times as the consummation or perfection of God's creation of the world. For example, according to some ancient Hebrew worldviews, reality unfolds along a linear path (or rather, a spiral path, with cyclical components that nonetheless have a linear trajectory); the world began with God and is ultimately headed toward God's final goal for creation, the world to come.

Eschatologies vary as to their degree of optimism or pessimism about the future. In some eschatologies, conditions are better for some and worse for others, e.g. "heaven and hell".

Eschatology in Religions

Bahá'í

In Bahá'í belief, creation has neither a beginning nor an end. Instead, the eschatology of other religions is viewed as symbolic. In Bahá'í belief, human time is marked by a series of progressive revelations in which successive messengers or prophets come from God. The coming of each of these messengers is seen as the day of judgment to the adherents of the previous religion, who may choose to accept the new messenger and enter the "heaven" of belief, or denounce the new messenger and enter the "hell" of denial. In this view, the terms "heaven" and "hell" are seen as symbolic terms for the person's spiritual progress and their nearness to or distance from God. In Bahá'í belief, the coming of Bahá'u'lláh, the founder of the Bahá'í Faith, signals the fulfilment of previous eschatological expectations of Islam, Christianity and other major religions.

Buddhism

Christianity

Christian eschatology is concerned with death, an intermediate state, Heaven, hell, the return of Jesus, and the resurrection of the dead. Several evangelical denominations include a rapture, a great tribulation, the Millennium, end of the world, the last judgment, a new heaven and a new earth (the World to Come), and the ultimate consummation of all of God's purposes. Eschatological passages are found in many places, esp. Isaiah, Daniel, Ezekiel, Matthew 24, *The Sheep and the Goats*, and the Book of Revelation, but Revelation often occupies a central place in Christian eschatology.

The Second Coming of Christ is the central event in Christian eschatology. Most Christians believe that death and suffering will continue to exist until Christ's return. There are, however, various views concerning the order and significance of other eschatological events.

The Book of Revelation is at the core of Christian eschatology. The study of Revelation is usually divided into four interpretative methodologies or hermeneutics. In the Futurist approach, Revelation is treated mostly as unfulfilled prophecy taking place in some yet undetermined future. This is the approach which most applies to eschatological studies. In the Preterist approach, Revelation is chiefly interpreted as having prophetic fulfillment in the past, principally, the events of the first century CE, such as the struggle of Christianity to survive the persecutions of the Roman Empire, the fall of Jerusalem in 70 CE, and the desecration of the Temple in the same year. In the Historicist approach, Revelation provides a broad view of history, and passages in Revelation are identified with major historical people and events. In the Idealist (or Spiritualist or Symbolic) approach, the events of Revelation are neither past nor future, but are purely symbolic, dealing with the ongoing struggle and ultimate triumph of good over evil.

Hinduism

Contemporary Hindu eschatology is linked in the Vaishnavite tradition to the figure of Kalki, the tenth and last avatar of Vishnu before the age draws to a close who will reincarnate as Shiva simultaneously dissolves and regenerates the universe.

Most Hindus believe that the current period is the Kali Yuga, the last of four *Yuga* that make up the current age. Each period has seen successive degeneration in the moral order, to the point that in the Kali Yuga quarrel and hypocrisy are the norm. In Hinduism, time is cyclic, consisting of cycles or "kalpas". Each kalpa lasts $4.1 - 8.2$ billion years, which is a period of one full day and night for Brahma, who in turn will live for 311 trillion, 40 billion years. The cycle of birth, growth, decay, and renewal at the individual level finds its echo in the cosmic order, yet is affected by vagaries of divine intervention in Vaishnavite belief. Some Shaivites hold the view that Shiva is incessantly destroying and creating the world.

After this larger cycle, all of creation will contract to a singularity and then again will expand from that single point, as the ages continue in a religious fractal pattern.

Islam

Islamic eschatology is documented in the sayings of the Prophet Muhammad, regarding the Signs of the Day of Judgement. The Prophet's sayings on the subject have been traditionally divided into Major and Minor Signs. He spoke about several Minor Signs of the approach of the Day of Judgment, including:

- Abu Hurairah reported that Muhammad said: "If you survive for a time you would certainly see people who would have whips in their hands like the tail of an ox. They would get up in the morning under the wrath of God and they would go into the evening with the anger of God."

- Abu Hurairah narrated that Muhammad said, "When honesty is lost, then wait for the Day of Judgment." It was asked, "How will honesty be lost, O Messenger of God?" He said, "When authority is given to those who do not deserve it, then wait for the Day of Judgment."

- 'Umar ibn al-Khattāb, in a long narration, relating to the questions of the angel Gabriel, reported: "Inform me when the Day of Judgment will be." He [the Prophet Muhammad] remarked: "The one who is being asked knows no more than the inquirer." He [the inquirer] said: "Tell me about its indications." He [the Prophet Muhammad] said: "That the slave-girl gives birth to her mistress and master, and that you would find barefooted, destitute shepherds of goats vying with one another in the construction of magnificent buildings."

- "Before the Day of Judgment there will be great liars, so beware of them."

- "When the most wicked member of a tribe becomes its ruler, and the most worthless member of a community becomes its leader, and a man is respected through fear of the evil he may do, and leadership is given to people who are unworthy of it, expect the Day of Judgment."

Regarding the Major Signs, a Companion of the Prophet narrated: "Once we were sitting together and talking amongst ourselves when the Prophet appeared. He asked us what it was we were discussing. We said it was the Day of Judgment. He said: 'It will not be called until ten signs have appeared: Smoke, Dajjal (the Antichrist), the creature (that will wound the people), the rising of the sun in the West, the Second Coming of Jesus, the emergence of Gog and Magog, and three sinkings (or cavings in of the earth): one in the East, another in the West and a third in the Arabian Peninsula.'"

Judaism

Jewish eschatology is concerned with events that will happen in the end of days, according to the Hebrew Bible and Jewish thought. This includes the ingathering of the exiled diaspora, the coming of Jewish Messiah, afterlife, and the revival of the dead Tsadikim.

In Judaism, end times are usually called the "end of days", a phrase that appears several times in the Tanakh. The idea of a messianic age has a prominent place in Jewish thought, and is incorporated as part of the end of days.

Judaism addresses the end times in the Book of Daniel and numerous other prophetic passages in the Hebrew scriptures, and also in the Talmud, particularly Tractate Avodah Zarah.

Zoroastrianism

Frashokereti is the Zoroastrian doctrine of a final renovation of the universe, when evil will be destroyed, and everything else will be then in perfect unity with God (Ahura Mazda). The doctrinal premises are (1) good will eventually prevail over evil; (2) creation was initially perfectly good, but was subsequently corrupted by evil; (3) the world will ultimately be restored to the perfection it had at the time of creation; (4) the "salvation for the individual depended on the sum of [that person's] thoughts, words and deeds, and there could be no intervention, whether compassionate or capricious, by any divine being to alter this." Thus, each human bears the responsibility for the fate of his own soul, and simultaneously shares in the responsibility for the fate of the world.

Eschatology Equivalents in Science and Philosophy

Futures Studies and Transhumanism

Researchers in futures studies and transhumanists investigate how the accelerating rate of scientific progress may lead to a "technological singularity" in the future that would profoundly and unpredictably change the course of human history, and result in *Homo sapiens* no longer being the dominant life form on Earth.

Astronomy

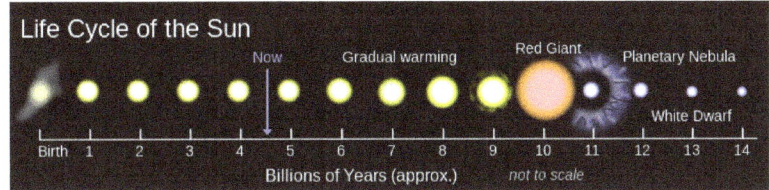

A diagram showing the life cycle of the Sun

Occasionally the term "physical eschatology" is applied to the long-term predictions of astrophysics. The Sun will turn into a red giant in approximately 5 billion years. Life on Earth will become impossible due to a rise in temperature long before the planet is actually swallowed up by the Sun.

Ultimate Fate of the Universe

The ultimate fate of the universe is a topic in physical cosmology, whose theoretical restrictions can usefully and scientifically predict the future behaviour of the universe as it ages. Based on available observational evidence, deciding the fate and evolution of the universe now have become valid cosmological questions, being beyond the mostly untestable constraints of mythological or theological beliefs. Many possible futures have been predicted by rival scientific hypotheses, including that the universe might have existed for a finite and infinite duration, or towards explaining how and in what circumstances it was created.

Observations made by Edwin Hubble during the 1920s-1930s found that most galaxies appeared to be moving away from each other, leading to current accepted Big Bang Theory. This suggests that the universe began in the far distant past about 13.8 billion years ago and ever since, continues to expand. Confirmation of the Big Bang mostly depends on knowing the rate of expansion, average density of matter, and the physical properties of the mass/energy in the universe.

There is a strong consensus among cosmologists that the universe is flat and will continue to expand forever. Yet many other factors may influence the universe's origin and final destiny, including, for example: the average motions of galaxies, the shape and structure of the universe, or the amount of dark matter and dark energy the universe contains.

Emerging Scientific Basis

Theory

The theoretical scientific exploration of the ultimate fate of the universe became possible with

Albert Einstein's 1916 theory of general relativity. General relativity can be employed to describe the universe on the largest possible scale. There are many possible solutions to the equations of general relativity, and each solution implies a possible ultimate fate of the universe.

Alexander Friedmann

Alexander Friedmann proposed several solutions in 1922, as did Georges Lemaître in 1927. In some of these solutions, the universe has been expanding from an initial singularity which was, essentially, the Big Bang.

Observation

In 1931, Edwin Hubble published his conclusion, based on his observations of Cepheid variable stars in distant galaxies, that the universe was expanding. From then on, the *beginning* of the universe and its possible *end* have been the subjects of serious scientific investigation.

Big Bang and Steady State Theories

In 1927, Georges Lemaître set out a theory that has since come to be called the Big Bang theory of the origin of the universe. In 1948, Fred Hoyle set out his opposing Steady State theory in which the universe continually expanded but remained statistically unchanged as new matter is constantly created. These two theories were active contenders until the 1965 discovery, by Arno Penzias and Robert Wilson, of the cosmic microwave background radiation, a fact that is a straightforward prediction of the Big Bang theory, and one that the original Steady State theory could not account for. As a result, The Big Bang theory quickly became the most widely held view of the origin of the universe.

Cosmological Constant

When Einstein formulated general relativity, he and his contemporaries believed in a static universe. When Einstein found that his equations could easily be solved in such a way as to allow the universe to be expanding now, and to contract in the far future, he added to those equations what he called a cosmological constant, essentially a constant energy density unaffected by any expansion or contraction, whose role was to offset the effect of gravity on the universe as a whole in such a way that the universe would remain static. After Hubble announced his conclusion that the universe was expanding, Einstein wrote that his cosmological constant was "the greatest blunder of my life".

Density Parameter

An important parameter in fate of the universe theory is the density parameter, Omega (Ω), defined as the average matter density of the universe divided by a critical value of that density. This selects one of three possible geometries depending on whether Ω is equal to, less than, or greater than 1. These are called, respectively, the flat, open and closed universes. These three adjectives refer to the overall geometry of the universe, and not to the local curving of spacetime caused by smaller clumps of mass (for example, galaxies and stars). If the primary content of the universe is inert matter, as in the dust models popular for much of the 20th century, there is a particular fate corresponding to each geometry. Hence cosmologists aimed to determine the fate of the universe by measuring Ω, or equivalently the rate at which the expansion was decelerating.

Repulsive Force

Starting in 1998, observations of supernovas in distant galaxies have been interpreted as consistent with a universe whose expansion is *accelerating*. Subsequent cosmological theorizing has been designed so as to allow for this possible acceleration, nearly always by invoking dark energy, which in its simplest form is just a positive cosmological constant. In general, dark energy is a catch-all term for any hypothesised field with negative pressure, usually with a density that changes as the universe expands.

Role of the Shape of the Universe

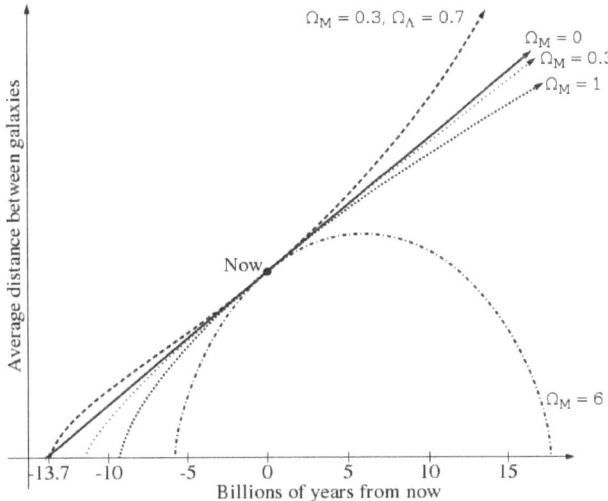

The ultimate fate of an expanding universe depends on the matter density Ω_M and the dark energy density Ω_Λ.

The current scientific consensus of most cosmologists is that the ultimate fate of the universe depends on its overall shape, how much dark energy it contains, and on the equation of state which determines how the dark energy density responds to the expansion of the universe. Recent observations conclude, from 7.5 billion years after the Big Bang, that the expansion rate of the universe has likely been increasing, commensurate with the Open Universe theory. However, other recent measurements by Wilkinson Microwave Anisotropy Probe suggest that the universe is either flat or very close to flat.

Closed Universe

If $\Omega > 1$, then the geometry of space is closed like the surface of a sphere. The sum of the angles of a triangle exceeds 180 degrees and there are no parallel lines; all lines eventually meet. The geometry of the universe is, at least on a very large scale, elliptic.

In a closed universe, gravity eventually stops the expansion of the universe, after which it starts to contract until all matter in the universe collapses to a point, a final singularity termed the "Big Crunch", the opposite of the Big Bang. Some new modern theories assume the universe may have a significant amount of dark energy, whose repulsive force may be sufficient to cause the expansion of the universe to continue forever—even if $\Omega > 1$.

Open Universe

If $\Omega < 1$, the geometry of space is open, i.e., negatively curved like the surface of a saddle. The angles of a triangle sum to less than 180 degrees, and lines that do not meet are never equidistant; they have a point of least distance and otherwise grow apart. The geometry of such a universe is hyperbolic.

Even without dark energy, a negatively curved universe expands forever, with gravity negligibly slowing the rate of expansion. With dark energy, the expansion not only continues but accelerates. The ultimate fate of an open universe is either universal heat death, the "Big Freeze", or the "Big Rip", where the acceleration caused by dark energy eventually becomes so strong that it completely overwhelms the effects of the gravitational, electromagnetic and strong binding forces.

Conversely, a *negative* cosmological constant, which would correspond to a negative energy density and positive pressure, would cause even an open universe to re-collapse to a big crunch. This option has been ruled out by observations.

Flat Universe

If the average density of the universe exactly equals the critical density so that $\Omega = 1$, then the geometry of the universe is flat: as in Euclidean geometry, the sum of the angles of a triangle is 180 degrees and parallel lines continuously maintain the same distance. Measurements from Wilkinson Microwave Anisotropy Probe have confirmed the universe is flat with only a 0.4% margin of error.

In absence of dark energy, a flat universe expands forever but at a continually decelerating rate, with expansion asymptotically approaching zero. With dark energy, the expansion rate of the universe initially slows down, due to the effect of gravity, but eventually increases. The ultimate fate of the universe is the same as an open universe.

Theories About the end of the Universe

The fate of the universe is determined by the density of the universe. The preponderance of evidence to date, based on measurements of the rate of expansion and the mass density, favors a universe that will continue to expand indefinitely, resulting in the "big freeze" scenario below. However, observations are not conclusive, and alternative models are still possible.

Big Freeze or Heat Death

The Big Freeze is a scenario under which continued expansion results in a universe that asymptotically approaches absolute zero temperature. This scenario, in combination with the Big Rip scenario, is currently gaining ground as the most important hypothesis. It could, in the absence of dark energy, occur only under a flat or hyperbolic geometry. With a positive cosmological constant, it could also occur in a closed universe. In this scenario, stars are expected to form normally for 10^{12} to 10^{14} (1–100 trillion) years, but eventually the supply of gas needed for star formation will be exhausted. As existing stars run out of fuel and cease to shine, the universe will slowly and inexorably grow darker. Eventually black holes will dominate the universe, which themselves will disappear over time as they emit Hawking radiation. Over infinite time, there would be a spontaneous entropy decrease by the Poincaré recurrence theorem, thermal fluctuations, and the fluctuation theorem.

A related scenario is heat death, which states that the universe goes to a state of maximum entropy in which everything is evenly distributed and there are no gradients—which are needed to sustain information processing, one form of which is life. The heat death scenario is compatible with any of the three spatial models, but requires that the universe reach an eventual temperature minimum.

Big Rip

In the special case of phantom dark energy, which has even more negative pressure than a simple cosmological constant, the density of dark energy increases with time, causing the *rate* of acceleration to increase, leading to a steady increase in the Hubble constant. As a result, all material objects in the universe, starting with galaxies and eventually (in a finite time) all forms, no matter how small, will disintegrate into unbound elementary particles and radiation, ripped apart by the phantom energy force and shooting apart from each other. The end state of the universe is a singularity, as the dark energy density and expansion rate becomes infinite.

Big Crunch

The Big Crunch hypothesis is a symmetric view of the ultimate fate of the universe. Just as the Big Bang started a cosmological expansion, this theory assumes that the average density of the universe is enough to stop its expansion and begin contracting. The end result is unknown; a simple estimation would have all the matter and space-time in the universe collapse into a dimensionless singularity, but at these scales unknown quantum effects need to be considered. Recent evidence suggests that this scenario is not likely but it has not been ruled out as measurements are only available over a short period of time and could reverse in the future.

This scenario allows the Big Bang to occur immediately after the Big Crunch of a preceding universe. If this happens repeatedly, it creates a cyclic model, which is also known as an oscillatory universe. The universe could then consist of an infinite sequence of finite universes, with each finite universe ending with a Big Crunch that is also the Big Bang of the next universe. Theoretically, the cyclic universe could not be reconciled with the second law of thermodynamics: entropy would build up from oscillation to oscillation and cause heat death. Current evidence also indicates the universe is not closed. This has caused cosmologists to abandon the oscillating universe model. A somewhat similar idea is embraced by the cyclic model, but this idea evades heat death because of an expansion of the branes that dilutes entropy accumulated in the previous cycle.

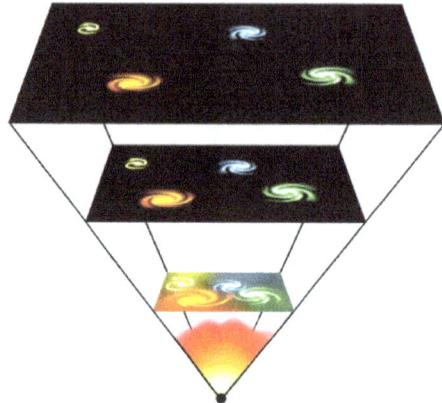

The Big Crunch. The vertical axis can be considered as either plus or minus time.

Big Bounce

The Big Bounce is a theorized scientific model related to the beginning of the known universe. It derives from the oscillatory universe or cyclic repetition interpretation of the Big Bang where the first cosmological event was the result of the collapse of a previous universe.

According to one version of the Big Bang theory of cosmology, in the beginning the universe was infinitely dense. Such a description seems to be at odds with everything else in physics, and especially quantum mechanics and its uncertainty principle. It is not surprising, therefore, that quantum mechanics has given rise to an alternative version of the Big Bang theory. Also, if the universe is closed, this theory would predict that once this universe collapses it will spawn another universe in an event similar to the Big Bang after a universal singularity is reached or a repulsive quantum force causes re-expansion.

In simple terms, this theory states that the universe will continuously repeat the cycle of a Big Bang, followed up with a Big Crunch.

False Vacuum

In order to best understand the false vacuum collapse theory, one must first understand the Higgs field which permeates the universe. Much like an electromagnetic field it varies in strength, based upon its potential. A true vacuum exists so long as the universe exists in its lowest energy state, in which case the false vacuum theory is irrelevant. However, if the vacuum is not in its lowest energy state (a false vacuum), it could tunnel into a lower energy state. This is called the vacuum metastability event. This has the potential to fundamentally alter our universe; in more audacious scenarios even the various physical constants could have different values, severely affecting the foundations of matter, energy, and spacetime. It is also possible that all structures will be destroyed instantaneously, without any forewarning. Studies of a particle similar to the Higgs boson support the theory of a false vacuum collapse billions of years from now.

According to the many-worlds interpretation of quantum mechanics, the universe will not end this way. Instead, each time a quantum event happens that causes the universe to decay from a false vacuum to a true vacuum state, the universe splits into several new worlds. In some of the new worlds the universe decays; in some others the universe continues as before.

Cosmic Uncertainty

Each possibility described so far is based on a very simple form for the dark energy equation of state. But as the name is meant to imply, very little is currently known about the physics of the dark energy. If the theory of inflation is true, the universe went through an episode dominated by a different form of dark energy in the first moments of the Big Bang; but inflation ended, indicating an equation of state far more complex than those assumed so far for present-day dark energy. It is possible that the dark energy equation of state could change again resulting in an event that would have consequences which are extremely difficult to predict or parametrize. As the nature of dark energy and dark matter remain enigmatic, even hypothetical, the possibilities surrounding their coming role in the universe are currently unknown.

Observational Constraints on Theories

Choosing among these rival scenarios is done by 'weighing' the universe, for example, measuring the relative contributions of matter, radiation, dark matter, and dark energy to the critical density. More concretely, competing scenarios are evaluated against data on galaxy clustering and distant supernovae, and on the anisotropies in the cosmic microwave background.

History of the Center of the Universe

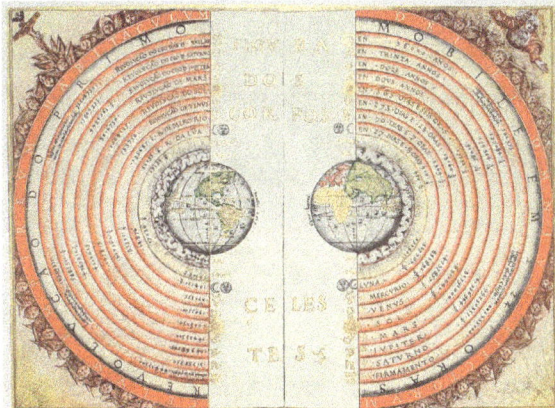

Figure of the heavenly bodies — An illustration of the Ptolemaic geocentric system by Portuguese cosmographer and cartographer Bartolomeu Velho, 1568 (Bibliothèque Nationale, Paris), depicting Earth as the centre of the Universe.

The center of the Universe is a concept that lacks a coherent definition in modern astronomy; according to standard cosmological theories on the shape of the universe, it has no center.

Historically, the center of the Universe had been believed to be a number of locations. Many mythological cosmologies included an *axis mundi*, the central axis of a flat Earth that connects the Earth, heavens, and other realms together. In the 4th century BCE Greece, the geocentric model was developed based on astronomical observation, proposing that the center of the Universe lies at the center of a spherical, stationary Earth, around which the sun, moon, planets, and stars rotate. With the development of the heliocentric model by Nicolaus Copernicus in the 16th century, the sun was believed to be the center of the Universe, with the planets (including Earth) and stars orbiting it.

In the early 20th century, the discovery of other galaxies and the development of the Big Bang theory led to the development of cosmological models of a homogeneous, isotropic Universe (which lacks a central point) that is expanding at all points.

Outside Astronomy

In religion or mythology, the *axis mundi* (also cosmic axis, world axis, world pillar, columna cerului, center of the world) is a point described as the center of the world, the connection between it and Heaven, or both.

Mount Hermon in Lebanon was regarded in some cultures as the *axis mundi*.

Mount Hermon was regarded as the axis mundi in Caananite tradition, from where the sons of God are introduced descending in 1 Enoch (1En6:6). The ancient Greeks regarded several sites as places of earth's *omphalos* (navel) stone, notably the oracle at Delphi, while still maintaining a belief in a cosmic world tree and in Mount Olympus as the abode of the gods. Judaism has the Temple Mount and Mount Sinai, Christianity has the Mount of Olives and Calvary, Islam has Mecca, said to be the place on earth that was created first, and the Temple Mount (Dome of the Rock). In Shinto, the Ise Shrine is the omphalos. In addition to the Kun Lun Mountains, where it is believed the peach tree of immortality is located, the Chinese folk religion recognizes four other specific mountains as pillars of the world.

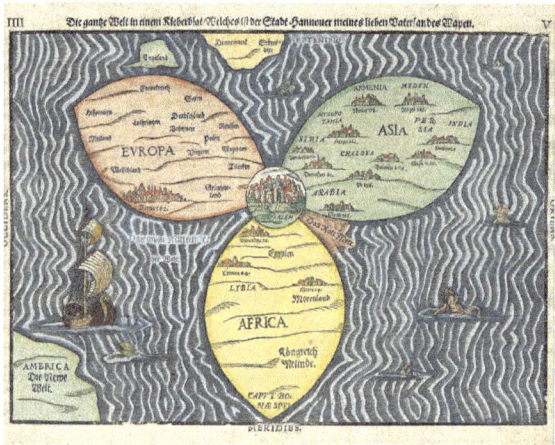

A 1581 map depicting Jerusalem as the center of the world.

Sacred places constitute world centers (omphalos) with the altar or place of prayer as the axis. Altars, incense sticks, candles and torches form the axis by sending a column of smoke, and prayer, toward heaven. The architecture of sacred places often reflects this role. "Every temple or palace--and by extension, every sacred city or royal residence--is a Sacred Mountain, thus becoming a Centre." The stupa of Hinduism, and later Buddhism, reflects Mount Meru. Cathedrals are laid out in the form of a cross, with the vertical bar representing the union of earth and heaven as the horizontal bars represent union of people to one another, with the altar at the intersection. Pagoda structures in Asian temples take the form of a stairway linking earth and heaven. A steeple in a church or a minaret in a mosque also serve as connections of earth and heaven. Structures such as the maypole, derived from the Saxons' Irminsul, and the totem pole among indigenous peoples of the Americas also represent world axes. The calumet, or sacred pipe, represents a column of smoke (the soul) rising form a world center. A mandala creates a world center within the boundaries of its two-dimensional space analogous to that created in three-dimensional space by a shrine.

In medieval times some Christians thought of Jerusalem as the center of the world (Latin: *umbilicus mundi*, Greek: *Omphalos*), and was so represented in the so-called T and O maps. Byzantine hymns speak of the Cross being "planted in the center of the earth."

Center of a Flat Earth

The Flammarion engraving (1888) depicts a traveller who arrives at the edge of a Flat Earth and sticks his head through the firmament.

The Flat Earth model is a belief that the Earth's shape is a plane or disk covered by a firmament contain heavenly bodies. Most pre-scientific cultures have had conceptions of a Flat Earth, including Greece until the classical period, the Bronze Age and Iron Age civilizations of the Near East until the Hellenistic period, India until the Gupta period (early centuries AD) and China until the 17th century. It was also typically held in the aboriginal cultures of the Americas, and a flat Earth domed by the firmament in the shape of an inverted bowl is common in pre-scientific societies.

"Center" is well-defined in a Flat Earth model. A flat earth would have a definite geographic center. There would also be a unique point at the exact center of a spherical firmament (or a firmament that was a half-sphere).

Earth as the Center of the Universe

The Flat Earth model gave way to an understanding of a Spherical Earth. Aristotle (384–322 BCE) provided observational arguments supporting the idea of a spherical Earth, namely that different stars are visible in different locations, travelers going south see southern constellations rise higher above the horizon, and the shadow of Earth on the Moon during a lunar eclipse is round, and spheres cast circular shadows while discs generally do not.

This understanding was accompanied by models of the Universe that depicted the Sun, Moon, stars, and naked eye planets circling the spherical Earth, including the noteworthy models of Aristotle and Ptolemy. This geocentric model was the dominant model from the 4th century BCE until the 17th century CE.

Sun as Center of the Universe

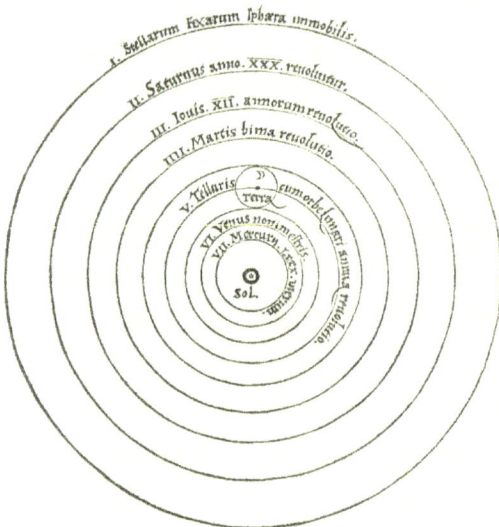

The heliocentric model from Nicolaus Copernicus' *De revolutionibus orbium coelestium*

Heliocentrism, or heliocentricism, is the astronomical model in which the Earth and planets revolve around a relatively stationary Sun at the center of our Solar System.

The notion that the Earth revolves around the Sun had been proposed as early as the 3rd century BCE by Aristarchus of Samos, but had received no support from most other ancient astronomers.

Nicolaus Copernicus' major theory of a heliocentric model was published in *De revolutionibus orbium coelestium* (*On the Revolutions of the Celestial Spheres*), in 1543, the year of his death, though he had formulated the theory several decades earlier. Copernicus' ideas were not immediately accepted, but they did begin a paradigm shift away from the Ptolemaic geocentric model to a heliocentric model. The Copernican revolution, as this paradigm shift would come to be called, would last until Isaac Newton's work over a century later.

Johannes Kepler published his first two laws about planetary motion in 1609, having found them by analyzing the astronomical observations of Tycho Brahe. Kepler's third law was published in 1619. The first law was "The orbit of every planet is an ellipse with the Sun at one of the two foci."

On 7 January 1610 Galileo used his telescope, with optics superior to what had been available before. He described "three fixed stars, totally invisible by their smallness", all close to Jupiter, and lying on a straight line through it. Observations on subsequent nights showed that the positions of these "stars" relative to Jupiter were changing in a way that would have been inexplicable if they had really been fixed stars. On 10 January Galileo noted that one of them had disappeared, an observation which he attributed to its being hidden behind Jupiter. Within a few days he concluded that they were orbiting Jupiter: Galileo stated that he had reached this conclusion on 11 January. He had discovered three of Jupiter's four largest satellites (moons). He discovered the fourth on 13 January.

His observations of the satellites of Jupiter created a revolution in astronomy: a planet with smaller planets orbiting it did not conform to the principles of Aristotelian Cosmology, which held that all heavenly bodies should circle the Earth, and many astronomers and philosophers initially refused to believe that Galileo could have discovered such a thing.

Newton made clear his heliocentric view of the solar system – developed in a somewhat modern way, because already in the mid-1680s he recognised the "deviation of the Sun" from the centre of gravity of the solar system. For Newton, it was not precisely the centre of the Sun or any other body that could be considered at rest, but rather "the common centre of gravity of the Earth, the Sun and all the Planets is to be esteem'd the Centre of the World", and this centre of gravity "either is at rest or moves uniformly forward in a right line" (Newton adopted the "at rest" alternative in view of common consent that the centre, wherever it was, was at rest).

Milky Way's Galactic Center as Center of the Universe

Great Andromeda Nebula by Isaac Roberts *(1899)*

In 1750 Thomas Wright, in his work *An original theory or new hypothesis of the Universe*, correctly speculated that the Milky Way might be a body of a huge number of stars held together by gravitational forces rotating about a Galactic Center, akin to the solar system but on a much larger scale. The resulting disk of stars can be seen as a band on the sky from our perspective inside the disk. In a treatise in 1755, Immanuel Kant elaborated on Wright's idea about the structure of the Milky Way. At the time, the existence of other galaxies had not been discovered.

The Nonexistence of a Center of the Universe

In 1917, Heber Doust Curtis observed a nova within what then was called the "Andromeda Nebula". Searching the photographic record, 11 more novae were discovered. Curtis noticed that novas in Andromeda were drastically fainter than novas in the Milky Way. Based on this, Curtis was able to estimate that Andromeda was 500,000 light-years away. As a result, Curtis became a proponent of the so-called "island Universes" hypothesis, which held that objects previously believed to be spiral nebulae within the Milky Way were actually independent galaxies.

In 1920, the Great Debate between Harlow Shapley and Curtis took place, concerning the nature of the Milky Way, spiral nebulae, and the dimensions of the Universe. To support his claim that the Great Andromeda Nebula (M31) was an external galaxy, Curtis also noted the appearance of dark lanes resembling the dust clouds in our own Galaxy, as well as the significant Doppler shift. In 1922 Ernst Öpik presented a very elegant and simple astrophysical method to estimate the distance of

M31. His result put the Andromeda Nebula far outside our Galaxy at a distance of about 450,000 parsec, which is about 1,500,000 ly. Edwin Hubble settled the debate about whether other galaxies exist in 1925 when he identified extragalactic Cepheid variable stars for the first time on astronomical photos of M31. These were made using the 2.5 metre (100 in) Hooker telescope, and they enabled the distance of Great Andromeda Nebula to be determined. His measurement demonstrated conclusively that this feature was not a cluster of stars and gas within our Galaxy, but an entirely separate galaxy located a significant distance from our own. This proved the existence of other galaxies.

Expanding Universe

Hubble also demonstrated that the redshift of other galaxies is approximately proportional to their distance from the Earth (Hubble's law). This raised the appearance of our galaxy being in the center of an expanding Universe, however, Hubble rejected the findings philosophically:

...if we see the nebulae all receding from our position in space, then every other observer, no matter where he may be located, will see the nebulae all receding from his position. However, the assumption is adopted. There must be no favoured location in the Universe, no centre, no boundary; all must see the Universe alike. And, in order to ensure this situation, the cosmologist, postulates spatial isotropy and spatial homogeneity, which is his way of stating that the Universe must be pretty much alike everywhere and in all directions."

The redshift observations of Hubble, in which galaxies appear to be moving away from us at a rate proportional to their distance from us, are now understood to be a result of the metric expansion of space. This is the increase of the distance between two distant parts of the Universe with time, and is an intrinsic expansion whereby the scale of space itself changes. As Hubble theorized, all observers anywhere in the Universe will observe a similar effect.

Copernican and Cosmological Principles

The Copernican principle, named after Nicolaus Copernicus, states that the Earth is not in a central, specially favored position. Hermann Bondi named the principle after Copernicus in the mid-20th century, although the principle itself dates back to the 16th-17th century paradigm shift away from the geocentric Ptolemaic system.

The cosmological principle is an extension of the Copernican principle which states that the Universe is homogeneous (the same observational evidence is available to observers at different locations in the Universe) and isotropic (the same observational evidence is available by looking in any direction in the Universe). A homogeneous, isotropic Universe does not have a center.

Axis Mundi

The axis mundi (also cosmic axis, world axis, world pillar, center of the world, world tree), in certain beliefs and philosophies, is the world center, or the connection between Heaven and Earth. As the celestial pole and geographic pole, it expresses a point of connection between sky and earth where the four compass directions meet. At this point travel and correspondence is made between higher and lower realms. Communication from lower realms may ascend to higher ones and blessings from higher realms may descend to lower ones and be disseminated to all. The spot functions as the *omphalos* (navel), the world's point of beginning.

Mount Kailash, depicting the holy family: Shiva and Parvati, cradling Skanda with Ganesha by Shiva's side

The image relates to the center of the earth (perhaps like an umbilical providing nourishment). It may have the form of a natural object (a mountain, a tree, a vine, a stalk, a column of smoke or fire) or a product of human manufacture (a staff, a tower, a ladder, a staircase, a maypole, a cross, a steeple, a rope, a totem pole, a pillar, a spire). Its proximity to heaven may carry implications that are chiefly religious (pagoda, temple mount, minaret, church) or secular (obelisk, lighthouse, rocket, skyscraper). The image appears in religious and secular contexts. The *axis mundi* symbol may be found in cultures utilizing shamanic practices or animist belief systems, in major world religions, and in technologically advanced "urban centers". In Mircea Eliade's opinion, "Every Microcosm, every inhabited region, has a Centre; that is to say, a place that is sacred above all." The axis mundi is often associated with mandalas.

Background

Mount Fuji, Japan

The symbol originates in a natural and universal psychological perception: that the spot one occupies stands at "the center of the world". This space serves as a microcosm of order because it is known and settled. Outside the boundaries of the microcosm lie foreign realms that, because they are unfamiliar or not ordered, represent chaos, death or night. From the center one may still venture in any of the four cardinal directions, make discoveries, and establish new centers as new realms become known and settled. The name of China, meaning "Middle Nation", is often interpreted as an expression of an ancient perception that the Chinese polity (or group of polities) occupied the center of the world, with other lands lying in various directions relative to it.

Within the central known universe a specific locale-often a mountain or other elevated place, a spot where earth and sky come closest gains status as center of the center, the axis mundi. High mountains are typically regarded as sacred by peoples living near them. Shrines are often erected at the summit or base. Mount Kunlun fills a similar role in China. For the ancient Hebrews Mount Zion expressed the symbol. Sioux beliefs take the Black Hills as the axis mundi. Mount Kailash is holy to Hinduism and several religions in Tibet. The Pitjantjatjara people in central Australia consider Uluru to be central to both their world and culture. In ancient Mesopotamia the cultures of ancient Sumer and Babylon erected artificial mountains, or ziggurats, on the flat river plain. These supported staircases leading to temples at the top. The Hindu temples in India are often situated on high mountains. E.g. Amarnath, Tirupati, Vaishno Devi etc. The pre-Columbian residents of Teotihuacán in Mexico erected huge pyramids featuring staircases leading to heaven. Jacob's Ladder is an axis mundi image, as is the Temple Mount. For Christians the Cross on Mount Calvary expresses the symbol. The Middle Kingdom, China, had a central mountain, Kunlun, known in Taoist literature as "the mountain at the middle of the world." To "go into the mountains" meant to dedicate oneself to a spiritual life. Monasteries of all faiths tend, like shrines, to be placed at elevated spots. Wise religious teachers are typically depicted in literature and art as bringing their revelations at world centers: mountains, trees, temples.

Yggdrasil, the World Ash in Norse myths

Because the axis mundi is an idea that unites a number of concrete images, no contradiction exists in regarding multiple spots as "the center of the world". The symbol can operate in a number of locales at once. Mount Hermon was regarded as the axis mundi in Caananite tradition, from where the sons of God are introduced descending in 1 Enoch (1En6:6). The ancient Greeks regarded several sites as places of earth's *omphalos* (navel) stone, notably the oracle at Delphi, while still maintaining a belief in a cosmic world tree and in Mount Olympus as the abode of the gods. Judaism has the Temple Mount and Mount Sinai, Christianity has the Mount of Olives and Calvary, Islam has Ka'aba, said to be the first building on earth, and the Temple Mount (Dome of the Rock). In Hinduism, Mount Kailash is identified with the mythical Mount Meru and regarded as the home of Shiva; in Vajrayana Buddhism, Mount Kailash is recognized as the most sacred place where all the dragon currents converge and is regarded as the gateway to Shambhala. In Shinto, the Ise Shrine is the omphalos. In addition to the Kunlun Mountains, where it is believed the peach tree of immortality is located, the Chinese folk religion recognizes four other specific mountains as pillars of the world.

Sacred places constitute world centers (omphalos) with the altar or place of prayer as the axis. Altars, incense sticks, candles and torches form the axis by sending a column of smoke, and prayer, toward heaven. The architecture of sacred places often reflects this role. "Every temple or palace--and by extension, every sacred city or royal residence--is a Sacred Mountain, thus becoming a Centre." The stupa of Hinduism, and later Buddhism, reflects Mount Meru. Cathedrals are laid out in the form of a cross, with the vertical bar representing the union of earth and heaven as the horizontal bars represent union of people to one another, with the altar at the intersection. Pagoda structures in Asian temples take the form of a stairway linking earth and heaven. A steeple in a church or a minaret in a mosque also serve as connections of earth and heaven. Structures such as the maypole, derived from the Saxons' Irminsul, and the totem pole among indigenous peoples of the Americas also represent world axes. The calumet, or sacred pipe, represents a column of smoke (the soul) rising form a world center. A mandala creates a world center within the boundaries of its two-dimensional space analogous to that created in three-dimensional space by a shrine.

Plants

Plants often serve as images of the axis mundi. The image of the Cosmic Tree provides an axis symbol that unites three planes: sky (branches), earth (trunk) and underworld (roots). In some Pacific island cultures the banyan tree, of which the Bodhi tree is of the Sacred Fig variety, is the abode of ancestor spirits. In Hindu religion, the banyan tree is considered sacred and is called *ashwath vriksha* ("I am banyan tree among trees" - *Bhagavad Gita*). It represents eternal life because of its seemingly ever-expanding branches. The Bodhi tree is also the name given to the tree under which Gautama Siddhartha, the historical Buddha, sat on the night he attained enlightenment. The Yggdrasil, or World Ash, functions in much the same way in Norse mythology; it is the site where Odin found enlightenment. Other examples include Jievaras in Lithuanian mythology and Thor's Oak in the myths of the pre-Christian Germanic peoples. The Tree of Life and the Tree of Knowledge of Good and Evil in Genesis present two aspects of the same image. Each is said to stand at the center of the Paradise garden from which four rivers flow to nourish the whole world. Each tree confers a boon. Bamboo, the plant from which Asian calligraphy pens are made, represents knowledge and is regularly found on Asian college campuses. The Christmas tree, which can be traced in its origins back to pre-Christian European beliefs, represents an axis mundi. Entheogens (psychoactive substances) are often regarded as world axes, such as the Fly Agaric mushroom among the Evenks of Russia. In China, traditional cosmography sometimes depicts the world center marked with the Jian tree. Two more trees are placed at the East and West, corresponding to the points of sunrise and sunset, as described in the *Huainanzi*. The Mesoamerican world tree connects the planes of the Underworld and the sky with that of the terrestrial realm.

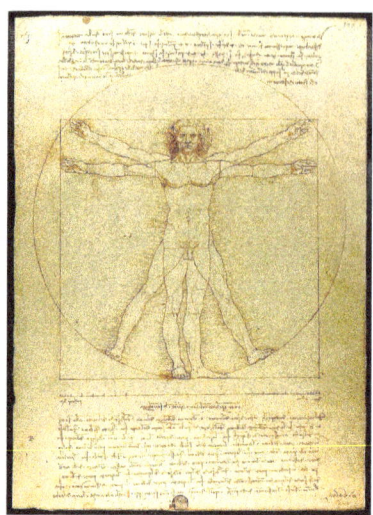

Vitruvian Man by Leonardo da Vinci (c. 1492)

Human Figure

The human body can express the symbol of world axis. Some of the more abstract Tree of Life representations, such as the *sefirot* in Kabbalism and in the *chakra* system recognized by Hinduism and Buddhism, merge with the concept of the human body as a pillar between heaven and earth. Disciplines such as yoga and tai chi begin from the premise of the human body as axis mundi. The Buddha represents a world centre in human form. Large statues of a meditating figure unite the human figure with the symbolism of temple and tower. Astrology in all its forms assumes a connection between human health and affairs and the orientation of these with celestial bodies. World religions regard the body itself as a temple and prayer as a column uniting earth to heaven. The ancient Colossus of Rhodes combined the role of human figure with those of portal and skyscraper. The image of a human being suspended on a tree or a cross locates the figure at the axis where heaven and earth meet. The Renaissance image known as the Vitruvian Man represented a symbolic and mathematical exploration of the human form as world axis.

Homes

The caduceus

Homes can represent world centers. The symbolism for their residents is the same as for inhabitants of palaces and other sacred mountains. The hearth participates in the symbolism of the altar and a central garden participates in the symbolism of primordial paradise. In Asian cultures houses were traditionally laid out in the form of a square oriented toward the four compass directions. A traditional Asian home was oriented toward the sky through feng shui, a system of geomancy, just as a palace would be. Traditional Arab houses are also laid out as a square surrounding a central fountain that evokes a primordial garden paradise. Mircea Eliade noted that "the symbolism of the pillar in [European] peasant houses likewise derives from the 'symbolic field' of the *axis mundi*. In many archaic dwellings the central pillar does in fact serve as a means of communication with the heavens, with the sky." The nomadic peoples of Mongolia and the Americas more often lived in circular structures. The central pole of the tent still operated as an axis but a fixed reference to the four compass points was avoided.

Shamanic Function

A common shamanic concept, and a universally told story, is that of the healer traversing the axis mundi to bring back knowledge from the other world. It may be seen in the stories from Odin and the World Ash Tree to the Garden of Eden and Jacob's Ladder to Jack and the Beanstalk and Rapunzel. It is the essence of the journey described in *The Divine Comedy* by Dante Alighieri. The epic poem relates its hero's descent and ascent through a series of spiral structures that take him from through the core of the earth, from the depths of Hell to celestial Paradise. It is also a central tenet in the Southeastern Ceremonial Complex.

Anyone or anything suspended on the axis between heaven and earth becomes a repository of potential knowledge. A special status accrues to the thing suspended: a serpent, a victim of crucifixion or hanging, a rod, a fruit, mistletoe. Derivations of this idea find form in the Rod of Asclepius, an emblem of the medical profession, and in the caduceus, an emblem of correspondence and commercial professions. The staff in these emblems represents the axis mundi while the serpents act as guardians of, or guides to, knowledge.

Modern Expressions

Taipei 101 (Taiwan)

Axis mundi symbolism continues to be evoked even in modern societies. The idea has proven especially consequential in the realm of architecture. Capitol buildings, as the direct descendents of palaces, fill this role, as do commemorative structures such as the Washington Monument in the United States. A skyscraper, as the term itself suggests, suggests the connection of earth and sky, as do spire structures of all sorts. Such buildings come to be regarded as "centers" of an inhabited area, or even the world, and serve as icons of its ideals. The first skyscraper of modern times, the Eiffel Tower, exemplifies this role. The structure was erected in 1889 in Paris, France, to serve as the centerpiece for the Exposition Universelle, making it a symbolic world center from the planning stages. It has served as an iconic image for the city and the nation ever since. Landmark skyscrapers often take names that clearly identify them as centers.

Designers of skyscrapers today routinely evoke the axis mundi symbolism inherent in ancient precedents. Taipei 101 in Taiwan, completed in 2004, evokes the staircase, bamboo stalk, pagoda, pillar and torch. The design of the Burj Khalifa (United Arab Emirates) evokes both desert plants and traditional Arab spires. William F. Baker, one of the designers, states that "the goal of the Burj Dubai [subsequently renamed Burj Khalifa] is not simply to be the world's tallest building--it is to embody the world's highest aspirations." Twin towers, such as the Petronas Towers (Kuala Lumpur, Malaysia) and the former World Trade Center (Manhattan), maintain the axis symbolism even as they more obviously assume the role of pillars. Some structures pierce the sky, implying movement or flight (Chicago Spire, CN Tower in Toronto, the Space Needle in Seattle). Some structures highlight the more lateral elements of the symbol in implying portals (Tuntex Sky Tower in Kaohsiung, Taiwan, The Gateway Arch in St. Louis).

The places with economic importance and where skyscrapers are founded are recognised as Financial centres. Examples of financial centres are London, New York City, Rome, Paris, Tokyo, Hong Kong, Chicago, Seoul, Shanghai, Toronto, Montreal, São Paulo, Frankfurt, and Amsterdam.

A geodesic place is another modern symbolism. Brasília, capital of Brazil, is known as a Geodesic place, where it is positioned at the middle of the country, on a drainage divide.

Ancient traditions continue in modern structures. The Peace Pagodas built since the 1947 unite religious and secular purposes in one symbol drawn from Buddhism. The influence of the pagoda tradition may be seen in modern Asian skyscrapers (Taipei 101, Petronas Towers). The ancient ziggurat has likewise reappeared in modern form, including the headquarters of the National Geographic Society in Washington, DC and The Ziggurat housing the California Department of General Services. Architect Frank Lloyd Wright conceived the Guggenheim Museum in New York as an inverted ziggurat. The Washington Monument is a modern obelisk.

Artistic representations of the world axis abound. Prominent among these is the *Colonne sans fin* (*The Endless Column*, 1938) an abstract sculpture by Romanian Constantin Brâncuşi. The column takes the form of a "sky pillar" (columna cerului) upholding the heavens even as its rhythmically repeating segments invite climb and suggest the possibility of ascension.

The association of the cosmic pillar with knowledge gives it a prominent role in the world of scholarship. University campuses typically assign a prominent axis role to a campus structure, such as a clock tower, library tower or bell tower. The building serves as the symbolic center of the settlement represented by the campus and serves as an emblem of its ideals. This symbolism of the center is closely tied to the widespread symbolism of the world axis. The image of the "ivory tower", a colloquial metaphor for academia, sustains the metaphor.

The image still takes natural forms as well, as in the American tradition of the Liberty Tree located at town centers. Individual homes continue to act as world axes, especially where Feng shui and other geomantic practices continue to be observed.

The corner of Haight and Ashbury Streets in San Francisco, California is regarded as the axis mundi in the hippie subculture. Christopher Street in Manhattan in New York City is the axis mundi in the gay subculture. Folsom Street, also in San Francisco, is the axis mundi in the leather subculture.

Axis mundi symbolism may be seen in much of the romance surrounding space travel. A rocket on the pad takes on all the symbolism of a tower and the astronaut enacts a mythic story. Each embarks on a perilous journey into the heavens and, if successful, returns with a boon for dissemination. The Apollo 13 insignia stated it succinctly: *Ex luna scientia* ("From the Moon, knowledge").

In fiction, Stephen King's 'The Dark Tower' series, and much of his other connected fiction, revolves around a tower that serves as the axis of all realities.

Geocentric Model

In astronomy, the geocentric model (also known as geocentrism, or the Ptolemaic system) is a superseded description of the universe with Earth at the centre. Under the geocentric model, the Sun, Moon, stars, and planets all circled Earth. The geocentric model served as the predominant description of the cosmos in many ancient civilizations, such as those of Aristotle and Ptolemy.

Two observations supported the idea that Earth was the center of the Universe. First, the Sun appears to revolve around Earth once per day. While the Moon and the planets have their own motions, they also appear to revolve around Earth about once per day. The stars appeared to be on a celestial sphere, rotating once each day along an axis through the north and south geographic poles of Earth. Second, Earth does not seem to move from the perspective of an Earth-bound observer; it appears to be solid, stable, and unmoving.

Ancient Greek, ancient Roman and medieval philosophers usually combined the geocentric model with a spherical Earth. It is not the same as the older flat Earth model implied in some mythology. The ancient Jewish Babylonian uranography pictured a flat Earth with a dome-shaped rigid canopy named firmament placed over it. However, the ancient Greeks believed that the motions of the planets were circular and not elliptical, a view that was not challenged in Western culture until the 17th century through the synthesis of theories by Copernicus and Kepler.

The astronomical predictions of Ptolemy's geocentric model were used to prepare astrological and astronomical charts for over 1500 years. The geocentric model held sway into the early modern age, but from the late 16th century onward, it was gradually superseded by the Heliocentric model of Copernicus, Galileo and Kepler. There was much resistance to the transition between these two theories. Christian theologians were reluctant to reject a theory that agreed with Bible passages (e.g. "Sun, stand you still upon Gibeon", Joshua 10:12). Others felt a new, unknown theory could not subvert an accepted consensus for geocentrism.

Ancient Greece

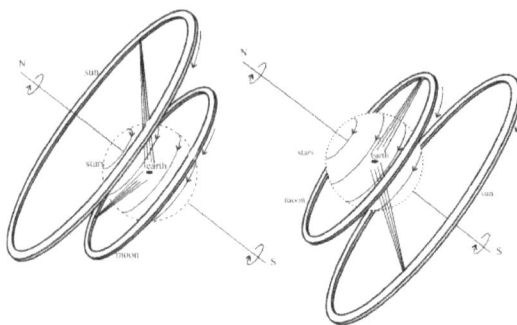

Illustration of Anaximander's models of the universe. On the left, daytime in summer;
on the right, nighttime in winter.

The geocentric model entered Greek astronomy and philosophy at an early point; it can be found in pre-Socratic philosophy. In the 6th century BC, Anaximander proposed a cosmology with Earth shaped like a section of a pillar (a cylinder), held aloft at the center of everything. The Sun, Moon, and planets were holes in invisible wheels surrounding Earth; through the holes, humans could see concealed fire. About the same time, Pythagoras thought that the Earth was a sphere (in accordance with observations of eclipses), but not at the center; they believed that it was in motion around an unseen fire. Later these views were combined, so most educated Greeks from the 4th century BC on thought that the Earth was a sphere at the center of the universe.

In the 4th century BC, two influential Greek philosophers, Plato and his student Aristotle, wrote works based on the geocentric model. According to Plato, the Earth was a sphere, stationary at the center of the universe. The stars and planets were carried around the Earth on spheres or circles, arranged in the order (outwards from the center): Moon, Sun, Venus, Mercury, Mars, Jupiter, Saturn, fixed stars, with the fixed stars located on the celestial sphere. In his "Myth of Er", a section of the *Republic*, Plato describes the cosmos as the Spindle of Necessity, attended by the Sirens and turned by the three Fates. Eudoxus of Cnidus, who worked with Plato, developed a less mythical, more mathematical explanation of the planets' motion based on Plato's dictum stating that all phenomena in the heavens can be explained with uniform circular motion. Aristotle elaborated on Eudoxus' system.

In the fully developed Aristotelian system, the spherical Earth is at the center of the universe, and all other heavenly bodies are attached to 47–55 transparent, rotating spheres surrounding the Earth, all concentric with it. (The number is so high because several spheres are needed for each planet.) These spheres, known as crystalline spheres, all moved at different uniform speeds to create the revolution of bodies around the Earth. They were composed of an incorruptible substance called aether. Aristotle believed that the moon was in the innermost sphere and therefore touches the realm of Earth, causing the dark spots (macula) and the ability to go through lunar phases. He further described his system by explaining the natural tendencies of the terrestrial elements: Earth, water, fire, air, as well as celestial aether. His system held that Earth was the heaviest element, with the strongest movement towards the center, thus water formed a layer surrounding the sphere of Earth. The tendency of air and fire, on the other hand, was to move upwards, away from the center, with fire being lighter than air. Beyond the layer of fire, were the solid spheres of aether in which the celestial bodies were embedded. They, themselves, were also entirely composed of aether.

Adherence to the geocentric model stemmed largely from several important observations. First of all, if the Earth did move, then one ought to be able to observe the shifting of the fixed stars due to stellar parallax. In short, if the Earth was moving, the shapes of the constellations should change considerably over the course of a year. If they did not appear to move, the stars are either much farther away than the Sun and the planets than previously conceived, making their motion undetectable, or in reality they are not moving at all. Because the stars were actually much further away than Greek astronomers postulated (making movement extremely subtle), stellar parallax was not detected until the 19th century. Therefore, the Greeks chose the simpler of the two explanations. The lack of any observable parallax was considered a fatal flaw in any non-geocentric theory. Another observation used in favor of the geocentric model at the time was the apparent consistency of Venus' luminosity, which implies that it is usually about the same distance from Earth, which in turn is more consistent with geocentrism than heliocentrism. In reality, that is because the loss of light caused by Venus' phases compensates for the increase in apparent size caused by its varying distance from Earth. Objectors to heliocentrism noted that terrestrial bodies naturally tend to come to rest as near as possible to the center of the Earth. Further barring the opportunity to fall closer the center, terrestrial bodies tend not to move unless forced by an outside object, or transformed to a different element by heat or moisture.

Atmospheric explanations for many phenomena were preferred because the Eudoxan–Aristotelian model based on perfectly concentric spheres was not intended to explain changes in the brightness of the planets due to a change in distance. Eventually, perfectly concentric spheres were abandoned as it was impossible to develop a sufficiently accurate model under that ideal. However, while providing for similar explanations, the later deferent and epicycle model was flexible enough to accommodate observations for many centuries.

Ptolemaic Model

Although the basic tenets of Greek geocentrism were established by the time of Aristotle, the details of his system did not become standard. The Ptolemaic system, developed by the Hellenistic astronomer Claudius Ptolemaeus in the 2nd century AD finally standardised geocentrism. His main astronomical work, the *Almagest*, was the culmination of centuries of work by Hellenic, Hellenistic and Babylonian astronomers. For over a millennium European and Islamic astronomers assumed it was the correct cosmological model. Because of its influence, people sometimes wrongly think the Ptolemaic system is identical with the geocentric model.

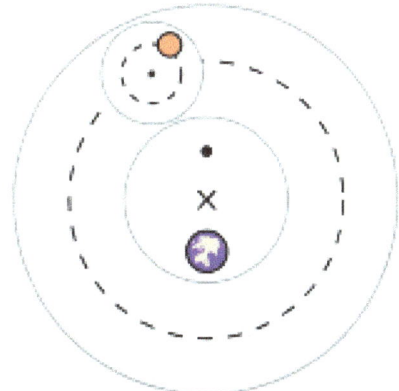

The basic elements of Ptolemaic astronomy, showing a planet on an epicycle with an eccentric deferent and an equant point. The Green shaded area is the celestial sphere which the planet occupies.

Ptolemy argued that the Earth was a sphere in the center of the universe, from the simple observation that half the stars were above the horizon and half were below the horizon at any time (stars on rotating stellar sphere), and the assumption that the stars were all at some modest distance from the center of the universe. If the Earth was substantially displaced from the center, this division into visible and invisible stars would not be equal.

Ptolemaic System

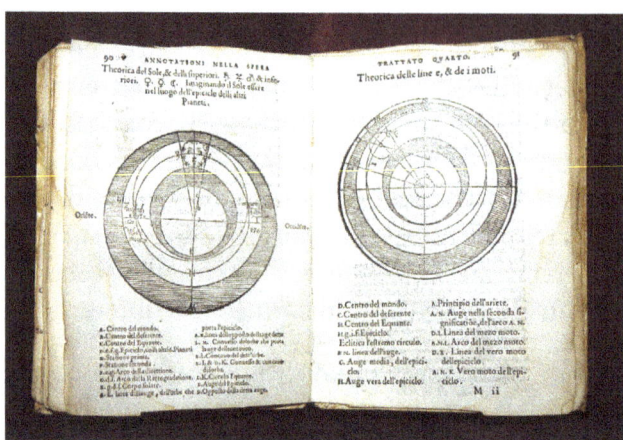

Pages from 1550 *Annotazione* on Sacrobosco's *Tractatus de Sphaera*, showing the Ptolemaic system.

In the Ptolemaic system, each planet is moved by a system of two spheres: one called its deferent; the other, its epicycle. The deferent is a circle whose center point, called the eccentric and marked in the diagram with an X, is removed from the Earth. The original purpose of the eccentric was to account for the differences of the lengths of the seasons (autumn is the shortest by a week or so), by placing the Earth away from the center of rotation of the rest of the universe. Another sphere, the epicycle, is embedded inside the deferent sphere and is represented by the smaller dotted line to the right. A given planet then moves around the epicycle at the same time the epicycle moves along the path marked by the deferent. These combined movements cause the given planet to move closer to and further away from the Earth at different points in its orbit, and explained the observation that planets slowed down, stopped, and moved backward in retrograde motion, and then again reversed to resume normal, or prograde, motion.

The deferent-and-epicycle model had been used by Greek astronomers for centuries along with the idea of the eccentric (a deferent which is slightly off-center from the Earth), which was even older. In the illustration, the center of the deferent is not the Earth but the spot marked X, making it eccentric, from which the spot takes its name. Unfortunately, the system that was available in Ptolemy's time did not quite match observations, even though it was considerably improved over Hipparchus' system. Most noticeably the size of a planet's retrograde loop (especially that of Mars) would be smaller, and sometimes larger, than expected, resulting in positional errors of as much as 30 degrees. To alleviate the problem, Ptolemy developed the equant. The equant was a point near the center of a planet's orbit which, if you were to stand there and watch, the center of the planet's epicycle would always appear to move at uniform speed; all other locations would see non-uniform speed, like on the Earth. By using an equant, Ptolemy claimed to keep motion which was uniform and circular, although it departed from the Platonic ideal of uniform circular motion. The resultant system, which eventually came to be widely accepted in the west, seems unwieldy to modern astronomers; each planet required an epicycle revolving on a deferent, offset by an equant which was different for each planet. It predicted various celestial motions, including the beginning and end of retrograde motion, to within a maximum error of 10 degrees, considerably better than without the equant.

The model with epicycles is in fact a very good model of an elliptical orbit with low eccentricity. The well known ellipse shape does not appear to a noticeable extent when the eccentricity is less than 5%, but the offset distance of the 'center' (in fact the focus occupied by the sun) is very noticeable even with low eccentricities as possessed by the planets.

To summarize, Ptolemy devised a system that was compatible with Aristotelian philosophy and managed to track actual observations and predict future movement mostly to within the limits of the next 1000 years of observations. The observed motions and his mechanisms for explaining them include:

The Ptolemaic System		
Object(s)	**Observation**	**Modeling mechanism**
Stars	Motion of entire sky E to W in ~24 hrs ("first motion")	Stars: Daily motion E to W of sphere of stars, carrying all other spheres with it. Normally ignored; other spheres have additional motions.
Sun	Motion yearly W to E along ecliptic	Motion of sun's sphere W to E in year
Sun	Non-uniform rate along ecliptic (uneven seasons)	Eccentric orbit (sun's deferent center off Earth)
Moon	Monthly motion W to E compared to stars	Monthly W to E motion of moon's sphere
The 5 planets	General motion W to E through zodiac	Motion of deferents W to E; period set by observation of planet going around the ecliptic.
Planets	Retrograde motion	Motion of epicycle in same direction as deferent. Period of epicycle is time between retrograde motions (synodic period).
Planets	Variations in speed through the zodiac	Eccentric per planet
Planets	Variations in retrograde timing	Equants per planet (Copernicus used a pair of epicycles instead)
Planets	Size of deferents, epicycles	Only ratio between radius of deferent and associated epicycle determined; absolute distances not determined in theory.
Interior planets	Average greatest elongations of 23 (Mercury) and 46 degrees (Venus)	Size of epicycles set by these angles, proportional to distances.
Interior planets	Limited to movement near the sun	Center their deferent centers along the Sun-Earth line.
Exterior planets	Retrograde only at opposition, when brightest	Radii of epicycles aligned to Sun-Earth line

The geocentric model was eventually replaced by the heliocentric model. The earliest heliocentric model, Copernican heliocentrism, could remove Ptolemy's epicycles because the retrograde motion could be seen to be the result of the combination of Earth and planet movement and speeds. Copernicus felt strongly that equants were a violation of Aristotelian purity, and proved that replacement of the equant with a pair of new epicycles was entirely equivalent. Astronomers often continued using the equants instead of the epicycles because the former was easier to calculate, and gave the same result.

It has been determined, in fact, that the Copernican, Ptolemaic and even the Tychonic models provided identical results to identical inputs. They are computationally equivalent. It wasn't until Kepler demonstrated a physical observation that could show that the physical sun is directly involved in determining an orbit that a new model was required.

The Ptolemaic order of spheres from Earth outward is:

1. Moon

2. Mercury

3. Venus

4. Sun

5. Mars

6. Jupiter

7. Saturn

8. Fixed Stars

9. Primum Mobile (First Moved).

Ptolemy did not invent or work out this order, which aligns with the ancient Seven Heavens religious cosmology common to the major Eurasian religious traditions. It also follows the decreasing orbital periods of the moon, sun, planets and stars.

Islamic Astronomy and Geocentrism

Muslim astronomers generally accepted the Ptolemaic system and the geocentric model, but by the 10th century texts appeared regularly whose subject matter was doubts concerning Ptolemy (*shukūk*). Several Muslim scholars questioned the Earth's apparent immobility and centrality within the universe. Some Muslim astronomers believed that the Earth rotates around its axis, such as Abu Sa'id al-Sijzi (d. circa 1020). According to al-Biruni, Sijzi invented an astrolabe called *al-zūraqī* based on a belief held by some of his contemporaries "that the motion we see is due to the Earth's movement and not to that of the sky." The prevalence of this view is further confirmed by a reference from the 13th century which states:

According to the geometers [or engineers] (*muhandisīn*), the Earth is in constant circular motion, and what appears to be the motion of the heavens is actually due to the motion of the Earth and not the stars.

Early in the 11th century Alhazen wrote a scathing critique of Ptolemy's model in his *Doubts on Ptolemy* (c. 1028), which some have interpreted to imply he was criticizing Ptolemy's geocentrism, but most agree that he was actually criticizing the details of Ptolemy's model rather than his geocentrism.

In the 12th century, Arzachel departed from the ancient Greek idea of uniform circular motions by hypothesizing that the planet Mercury moves in an elliptic orbit, while Alpetragius proposed a planetary model that abandoned the equant, epicycle and eccentric mechanisms, though this resulted in a system that was mathematically less accurate. Alpetragius also declared the Ptolemaic system as an imaginary model that was successful at predicting planetary positions but not real or physical. His alternative system spread through most of Europe during the 13th century.

Fakhr al-Din al-Razi (1149–1209), in dealing with his conception of physics and the physical world in his *Matalib*, rejects the Aristotelian and Avicennian notion of the Earth's centrality within the universe, but instead argues that there are "a thousand thousand worlds (*alfa alfi 'awalim*) beyond this world such that each one of those worlds be bigger and more massive than this world as well as having the like of what this world has." To support his theological argument, he cites the Qur'anic verse, "All praise belongs to God, Lord of the Worlds," emphasizing the term "Worlds."

The "Maragha Revolution" refers to the Maragha school's revolution against Ptolemaic astronomy. The "Maragha school" was an astronomical tradition beginning in the Maragha observatory and continuing with astronomers from the Damascus mosque and Samarkand observatory. Like their Andalusian predecessors, the Maragha astronomers attempted to solve the equant problem (the circle around whose circumference a planet or the center of an epicycle was conceived to move uniformly) and produce alternative configurations to the Ptolemaic model without abandoning geocentrism. They were more successful than their Andalusian predecessors in producing non-Ptolemaic configurations which eliminated the equant and eccentrics, were more accurate than the Ptolemaic model in numerically predicting planetary positions, and were in better agreement with empirical observations. The most important of the Maragha astronomers included Mo'ayyeduddin Urdi (d. 1266), Nasīr al-Dīn al-Tūsī (1201–1274), Qutb al-Din al-Shirazi (1236–1311), Ibn al-Shatir (1304–1375), Ali Qushji (c. 1474), Al-Birjandi (d. 1525), and Shams al-Din al-Khafri (d. 1550). Ibn al-Shatir, the Damascene astronomer (1304–1375 AD) working at the Umayyad Mosque, wrote a major book entitled *Kitab Nihayat al-Sul fi Tashih al-Usul* (*A Final Inquiry Concerning the Rectification of Planetary Theory*) on a theory which departs largely from the Ptolemaic system known at that time. In his book, *Ibn al-Shatir, an Arab astronomer of the fourteenth century*, E. S. Kennedy wrote "what is of most interest, however, is that Ibn al-Shatir's lunar theory, except for trivial differences in parameters, is identical with that of Copernicus (1473–1543 AD)." The discovery that the models of Ibn al-Shatir are mathematically identical to those of Copernicus suggests the possible transmission of these models to Europe. At the Maragha and Samarkand observatories, the Earth's rotation was discussed by al-Tusi and Ali Qushji (b. 1403); the arguments and evidence they used resemble those used by Copernicus to support the Earth's motion.

However, the Maragha school never made the paradigm shift to heliocentrism. The influence of the Maragha school on Copernicus remains speculative, since there is no documentary evidence to prove it. The possibility that Copernicus independently developed the Tusi couple remains open, since no researcher has yet demonstrated that he knew about Tusi's work or that of the Maragha school.

Geocentrism and Rival Systems

Not all Greeks agreed with the geocentric model. The Pythagorean system has already been mentioned; some Pythagoreans believed the Earth to be one of several planets going around a central fire. Hicetas and Ecphantus, two Pythagoreans of the 5th century BC, and Heraclides Ponticus in the 4th century BC, believed that the Earth rotated on its axis but remained at the center of the universe. Such a system still qualifies as geocentric. It was revived in the Middle Ages by Jean Buridan. Heraclides Ponticus was once thought to have proposed that both Venus and Mercury went around the Sun rather than the Earth, but this is no longer accepted. Martianus Capella definitely put Mercury and Venus in orbit around the Sun. Aristarchus of Samos was the most radical. He wrote a work, which has not survived, on heliocentrism, saying that the Sun was at the center of the universe, while the Earth and other planets revolved around it. His theory was not popular, and he had one named follower, Seleucus of Seleucia.

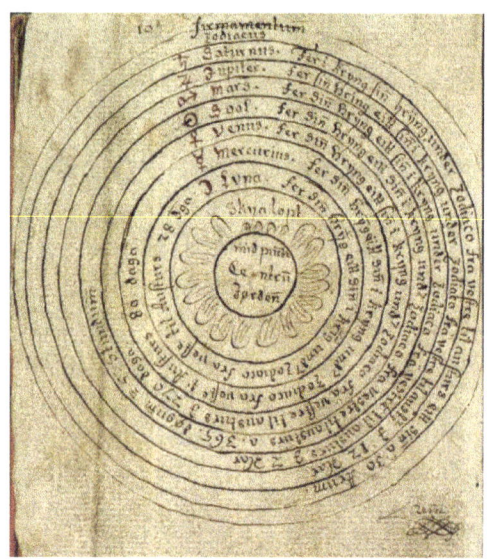

This drawing from an Icelandic manuscript dated around 1750 illustrates the geocentric model.

Copernican System

In 1543, the geocentric system met its first serious challenge with the publication of Copernicus' *De revolutionibus orbium coelestium* (*On the Revolutions of the Heavenly Spheres*), which posited that the Earth and the other planets instead revolved around the Sun. The geocentric system was still held for many years afterwards, as at the time the Copernican system did not offer better predictions than the geocentric system, and it posed problems for both natural philosophy and scripture. The Copernican system was no more accurate than Ptolemy's system, because it still used circular orbits. This was not altered until Johannes Kepler postulated that they were elliptical (Kepler's first law of planetary motion).

With the invention of the telescope in 1609, observations made by Galileo Galilei (such as that Jupiter has moons) called into question some of the tenets of geocentrism but did not seriously threaten it. Because he observed dark "spots" on the moon, craters, he remarked that the moon was not a perfect celestial body as had been previously conceived. This was the first time someone could see imperfections on a celestial body that was supposed to be composed of perfect aether. As such, because the moon's imperfections could now be related to those seen on Earth, one could argue that neither was unique: rather, they were both just celestial bodies made from Earth-like material. Galileo could also see the moons of Jupiter, which he dedicated to Cosimo II de' Medici, and stated that they orbited around Jupiter, not Earth. This was a significant claim as it would mean not only that not everything revolved around Earth as stated in the Ptolemaic model, but also showed a secondary celestial body could orbit a moving celestial body, strengthening the heliocentric argument that a moving Earth could retain the Moon. Galileo's observations were verified by other astronomers of the time period who quickly adopted use of the telescope, including Christoph Scheiner, Johannes Kepler, and Giovan Paulo Lembo.

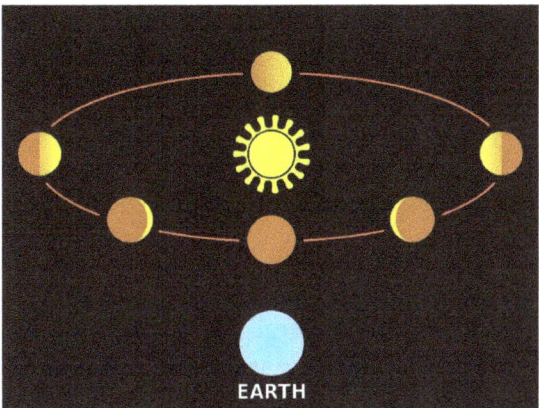

Phases of Venus

In December 1610, Galileo Galilei used his telescope to observe that Venus showed all phases, just like the Moon. He thought that while this observation was incompatible with the Ptolemaic system, it was a natural consequence of the heliocentric system.

However, Ptolemy placed Venus' deferent and epicycle entirely inside the sphere of the Sun (between the Sun and Mercury), but this was arbitrary; he could just as easily have swapped Venus and Mercury and put them on the other side of the Sun, or made any other arrangement of Venus and Mercury, as long as they were a lways near a line running from the Earth through the Sun, such as placing the center of the Venus epicycle near the Sun. In this case, if the Sun is the source of all the light, under the Ptolemaic system:

If Venus is between Earth and the Sun, the phase of Venus must always be crescent or all dark.

If Venus is beyond the Sun, the phase of Venus must always be gibbous or full.

But Galileo saw Venus at first small and full, and later large and crescent.

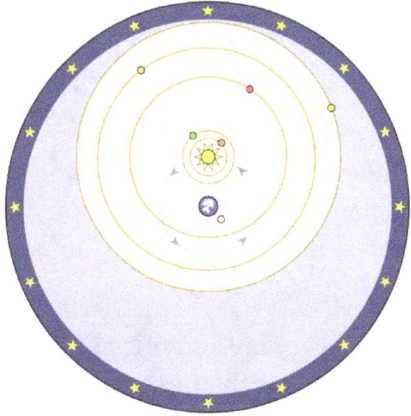

In this depiction of the Tychonic system, the objects on blue orbits (the moon and the sun)
revolve around the Earth. The objects on orange orbits (Mercury, Venus, Mars, Jupiter,
and Saturn) revolve around the sun. Around all is a sphere of stars, which rotates.

This showed that with a Ptolemaic cosmology, the Venus epicycle can be neither completely inside nor completely outside of the orbit of the Sun. As a result, Ptolemaics abandoned the idea that the epicycle of Venus was completely inside the Sun, and later 17th century competition between astronomical cosmologies focused on variations of Tycho Brahe's Tychonic system (in which the Earth was still at the center of the universe, and around it revolved the Sun, but all other planets revolved around the Sun in one massive set of epicycles), or variations on the Copernican system.

Gravitation

Johannes Kepler analysed Tycho Brahe's famously accurate observations and afterwards constructed his three laws in 1609 and 1619, based on a heliocentric view where the planets move in elliptical paths. Using these laws, he was the first astronomer to successfully predict a transit of Venus (for the year 1631). The change from circular orbits to elliptical planetary paths dramatically improved the accuracy of celestial observations and predictions. Because the heliocentric model by Copernicus was no more accurate than Ptolemy's system, new observations were needed to persuade those who still held on to the geocentric model. However, Kepler's laws based on Brahe's data became a problem which geocentrists could not easily overcome.

In 1687, Isaac Newton stated the law of universal gravitation, described earlier as a hypothesis by Robert Hooke and others. His main achievement was to mathematically derive Kepler's laws of planetary motion from the law of gravitation, thus helping to prove the latter. This introduced gravitation as the force that both kept the Earth and planets moving through the heavens and also kept the air from flying away. The theory of gravity allowed scientists to construct a plausible heliocentric model for the solar system quickly. In his *Principia*, Newton explained his system of how gravity, (previously thought of as an occult force) directed the movements of celestial bodies, and kept our solar system in its working order. His descriptions of centripetal force were a breakthrough in scientific thought which used the newly developed differential calculus, and finally replaced the previous schools of scientific thought, i.e. those of Aristotle and Ptolemy. However, the process was gradual.

Several empirical tests of Newton's theory, explaining the longer period of oscillation of a pendulum at the equator and the differing size of a degree of latitude, gradually became available over the period 1673–1738. In addition, stellar aberration was observed by Robert Hooke in 1674 and tested in a series of observations by Jean Picard over ten years finishing in 1680. However, it was not explained until 1729 when James Bradley provided an approximate explanation in terms of the Earth's revolution about the sun.

In 1838, astronomer Friedrich Wilhelm Bessel measured the parallax of the star 61 Cygni successfully, and disproved Ptolemy's claim that parallax motion did not exist. This finally confirmed the assumptions made by Copernicus, provided accurate, dependable scientific observations, and displayed truly how far away stars were from Earth.

A geocentric frame is useful for many everyday activities and most laboratory experiments, but is a less appropriate choice for solar-system mechanics and space travel. While a heliocentric frame is most useful in those cases, galactic and extra-galactic astronomy is easier if the sun is treated as neither stationary nor the center of the universe, but rotating around the center of our galaxy, and in turn our galaxy is also not at rest in the cosmic background.

Relativity

Albert Einstein and Leopold Infeld wrote in The Evolution of Physics (1938): "Can we formulate physical laws so that they are valid for all CS (=coordinate systems), not only those moving uniformly, but also those moving quite arbitrarily, relative to each other? If this can be done, our difficulties will be over. We shall then be able to apply the laws of nature to any CS. The struggle, so violent in the early days of science, between the views of Ptolemy and Copernicus would then be quite meaningless. Either CS could be used with equal justification. The two sentences, "the sun is at rest and the Earth moves", or "the sun moves and the Earth is at rest", would simply mean two different conventions concerning two different CS. Could we build a real relativistic physics valid in all CS; a physics in which there would be no place for absolute, but only for relative, motion? This is indeed possible!"

Religious and Contemporary Adherence to Geocentrism

The Ptolemaic model of the solar system held sway into the early modern age; from the late 16th century onward it was gradually replaced as the consensus description by the heliocentric model. Geocentrism as a separate religious belief, however, never completely died out. In the United States between 1870 and 1920, for example, various members of the Lutheran Church – Missouri Synod published articles disparaging Copernican astronomy, and geocentrism was widely taught within the synod during that period. However, in the 1902 *Theological Quarterly*, A. L. Graebner claimed that the synod had no doctrinal position on geocentrism, heliocentrism, or any scientific model, unless it were to contradict Scripture. He stated that any possible declarations of geocentrists within the synod did not set the position of the church body as a whole.

Articles arguing that geocentrism was the biblical perspective appeared in some early creation science newsletters pointing to some passages in the Bible, which, when taken literally, indicate that the daily apparent motions of the Sun and the Moon are due to their actual motions around the Earth rather than due to the rotation of the Earth about its axis. For example, in Joshua 10:12, the Sun and Moon are said to stop in the sky, and in Psalms the world is described as immobile. Psalms 93:1 says in part "the world is established, firm and secure".) Contemporary advocates for such religious beliefs include Robert Sungenis (president of Bellarmine Theological Forum and author of the 2006 book *Galileo Was Wrong*). These people subscribe to the view that a plain reading of the Bible contains an accurate account of the manner in which the universe was created and requires a geocentric worldview. Most contemporary creationist organizations reject such perspectives.

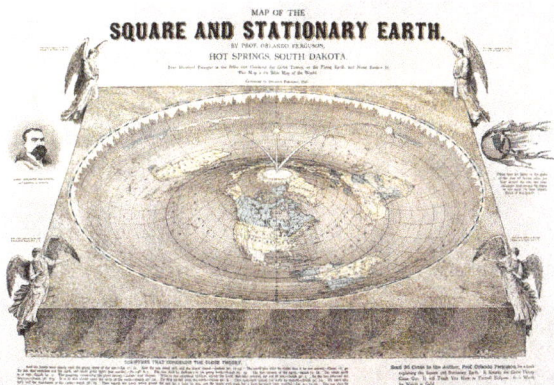

Map of the Square and Stationary Earth, by Orlando Ferguson (1893)

After all, Copernicanism was the first major victory of science over religion, so it's inevitable that some folks would think that everything that's wrong with the world began there. (Steven Dutch of the University of Wisconsin–Madison)

Morris Berman quotes a 2006 survey that show currently some 20% of the U.S. population believe that the sun goes around the Earth (geocentricism) rather than the Earth goes around the sun (heliocentricism), while a further 9% claimed not to know. Polls conducted by Gallup in the 1990s found that 16% of Germans, 18% of Americans and 19% of Britons hold that the Sun revolves around the Earth. A study conducted in 2005 by Jon D. Miller of Northwestern University, an expert in the public understanding of science and technology, found that about 20%, or one in five, of American adults believe that the Sun orbits the Earth. According to 2011 VTSIOM poll, 32% of Russians believe that the Sun orbits the Earth.

Historical Positions of the Roman Catholic Hierarchy

The famous Galileo affair pitted the geocentric model against the claims of Galileo. In regards to the theological basis for such an argument, two Popes addressed the question of whether the use of phenomenological language would compel one to admit an error in Scripture. Both taught that it would not. Pope Leo XIII (1878–1903) wrote:

"We have to contend against those who, making an evil use of physical science, minutely scrutinize the Sacred Book in order to detect the writers in a mistake, and to take occasion to vilify its contents... There can never, indeed, be any real discrepancy between the theologian and the physicist, as long as each confines himself within his own lines, and both are careful, as St. Augustine warns us, "not to make rash assertions, or to assert what is not known as known." If dissension should arise between them, here is the rule also laid down by St. Augustine, for the theologian: "Whatever they can really demonstrate to be true of physical nature, we must show to be capable of reconciliation with our Scriptures; and whatever they assert in their treatises which is contrary to these Scriptures of ours, that is to Catholic faith, we must either prove it as well as we can to be entirely false, or at all events we must, without the smallest hesitation, believe it to be so." To understand how just is the rule here formulated we must remember, first, that the sacred writers, or to speak more accurately, the Holy Ghost "Who spoke by them, did not intend to teach men these things (that is to say, the essential nature of the things of the visible universe), things in no way profitable unto salvation." Hence they did not seek to penetrate the secrets of nature, but rather described and dealt with things in more or less figurative language, or in terms which were commonly used at the time, and which in many instances are in daily use at this day, even by the most eminent men of science. Ordinary speech primarily and properly describes what comes under the senses; and somewhat in the same way the sacred writers-as the Angelic Doctor also reminds us – `went by what sensibly appeared," or put down what God, speaking to men, signified, in the way men could understand and were accustomed to. (Providentissimus Deus 18).

Maurice Finocchiaro, author of a book on the Galileo affair, notes that this is "a view of the relationship between biblical interpretation and scientific investigation that corresponds to the one advanced by Galileo in the "Letter to the Grand Duchess Christina". Pope Pius XII (1939–1958) repeated his predecessor's teaching:

The first and greatest care of Leo XIII was to set forth the teaching on the truth of the Sacred Books and to defend it from attack. Hence with grave words did he proclaim that there is no error whatsoever if the sacred writer, speaking of things of the physical order "went by what sensibly appeared" as the Angelic Doctor says, speaking either "in figurative language, or in terms which were commonly used at the time, and which in many instances are in daily use at this day, even among the most eminent men of science." For "the sacred writers, or to speak more accurately – the words are St. Augustine's – the Holy Spirit, Who spoke by them, did not intend to teach men these things – that is the essential nature of the things of the universe – things in no way profitable to salvation"; which principle "will apply to cognate sciences, and especially to history," that is, by refuting, "in a somewhat similar way the fallacies of the adversaries and defending the historical truth of Sacred Scripture from their attacks (*Divino afflante Spiritu*, 3).

In 1664 Pope Alexander VII republished the *Index Librorum Prohibitorum* (*List of Prohibited Books*) and attached the various decrees connected with those books, including those concerned with heliocentrism. He stated in a Papal Bull that his purpose in doing so was that "the succession of things done from the beginning might be made known [*quo rei ab initio gestae series innotescat*]."

The position of the curia evolved slowly over the centuries towards permitting the heliocentric view. In 1757, during the papacy of Benedict XIV, the Congregation of the Index withdrew the decree which prohibited *all* books teaching the Earth's motion, although the *Dialogue* and a few other books continued to be explicitly included. In 1820, the Congregation of the Holy Office, with the pope's approval, decreed that Catholic astronomer Giuseppe Settele was allowed to treat the Earth's motion as an established fact and removed any obstacle for Catholics to hold to the motion of the Earth:

The Assessor of the Holy Office has referred the request of Giuseppe Settele, Professor of Optics and Astronomy at La Sapienza University, regarding permission to publish his work Elements of Astronomy in which he espouses the common opinion of the astronomers of our time regarding the Earth's daily and yearly motions, to His Holiness through Divine Providence, Pope Pius VII. Previously, His Holiness had referred this request to the Supreme Sacred Congregation and concurrently to the consideration of the Most Eminent and Most Reverend General Cardinal Inquisitor. His Holiness has decreed that no obstacles exist for those who sustain Copernicus' affirmation regarding the Earth's movement in the manner in which it is affirmed today, even by Catholic authors. He has, moreover, suggested the insertion of several notations into this work, aimed at demonstrating that the above mentioned affirmation [of Copernicus], as it is has come to be understood, does not present any difficulties; difficulties that existed in times past, prior to the subsequent astronomical observations that have now occurred. [Pope Pius VII] has also recommended that the implementation [of these decisions] be given to the Cardinal Secretary of the Supreme Sacred Congregation and Master of the Sacred Apostolic Palace. He is now appointed the task of bringing to an end any concerns and criticisms regarding the printing of this book, and, at the same time, ensuring that in the future, regarding the publication of such works, permission is sought from the Cardinal Vicar whose signature will not be given without the authorization of the Superior of his Order.

In 1822, the Congregation of the Holy Office removed the prohibition on the publication of books treating of the Earth's motion in accordance with modern astronomy and Pope Pius VII ratified the decision:

The most excellent [cardinals] have decreed that there must be no denial, by the present or by future Masters of the Sacred Apostolic Palace, of permission to print and to publish works which treat of the mobility of the Earth and of the immobility of the sun, according to the common opinion of modern astronomers, as long as there are no other contrary indications, on the basis of the decrees of the Sacred Congregation of the Index of 1757 and of this Supreme [Holy Office] of 1820; and that those who would show themselves to be reluctant or would disobey, should be forced under punishments at the choice of [this] Sacred Congregation, with derogation of [their] claimed privileges, where necessary.

The 1835 edition of the Catholic Index of Prohibited Books for the first time omits the *Dialogue* from the list. In his 1921 papal encyclical, *In praeclara summorum*, Pope Benedict XV stated that, "though this Earth on which we live may not be the center of the universe as at one time was thought, it was the scene of the original happiness of our first ancestors, witness of their unhappy fall, as too of the Redemption of mankind through the Passion and Death of Jesus Christ." In 1965 the Second Vatican Council stated that, "Consequently, we cannot but deplore certain habits of mind, which are sometimes found too among Christians, which do not sufficiently attend to the rightful independence of science and which, from the arguments and controversies they spark, lead many minds to conclude that faith and science are mutually opposed." The footnote on this statement is to Msgr. Pio Paschini's, *Vita e opere di Galileo Galilei*, 2 volumes, Vatican Press (1964). Pope John Paul II regretted the treatment which Galileo received, in a speech to the Pontifical Academy of Sciences in 1992. The Pope declared the incident to be based on a "tragic mutual miscomprehension". He further stated:

Cardinal Poupard has also reminded us that the sentence of 1633 was not irreformable, and that the debate which had not ceased to evolve thereafter, was closed in 1820 with the imprimatur given to the work of Canon Settele. . . . The error of the theologians of the time, when they maintained the centrality of the Earth, was to think that our understanding of the physical world's structure was, in some way, imposed by the literal sense of Sacred Scripture. Let us recall the celebrated saying attributed to Baronius "Spiritui Sancto mentem fuisse nos docere quomodo ad coelum eatur, non quomodo coelum gradiatur". In fact, the Bible does not concern itself with the details of the physical world, the understanding of which is the competence of human experience and reasoning. There exist two realms of knowledge, one which has its source in Revelation and one which reason can discover by its own power. To the latter belong especially the experimental sciences and philosophy. The distinction between the two realms of knowledge ought not to be understood as opposition.

Orthodox Judaism

A few Orthodox Jewish leaders, particularly the Lubavitcher Rebbe, maintain a geocentric model of the universe based on the aforementioned Biblical verses and an interpretation of Maimonides to the effect that he ruled that the Earth is orbited by the sun. The Lubavitcher Rebbe also explained that geocentrism is defensible based on the theory of Relativity, which establishes that "when two bodies in space are in motion relative to one another, ... science declares with absolute certainty that from the scientific point of view both possibilities are equally valid, namely that the Earth revolves around the sun, or the sun revolves around the Earth."

The Zohar implies: "The entire world and those upon it, spin round in a circle like a ball,' both those at the bottom of the ball and those at the top. All God's creatures, wherever they live on the different parts of the ball, look different (in color, in their features) because the air is different in each place, but they stand erect as all other human beings.

Therefore there are places in the world where, when some have light, others have darkness; when some have day, others have night.

While geocentrism is important in Maimonides' calendar calculations, the great majority of Jewish religious scholars, who accept the divinity of the Bible and accept many of his rulings as legally binding, do not believe that the Bible or Maimonides command a belief in geocentrism.

Islam

Prominent cases of modern geocentrism in Islam are very isolated. Very few individuals promoted a geocentric view of the universe. One of them was Ahmed Raza Khan Barelvi, a Sunni scholar of Indian subcontinent. He rejected the heliocentric model and wrote a book that explains the movement of sun, moon and other planets around the Earth. The Grand Mufti of Saudi Arabia from 1993 to 1999, Ibn Baz also promoted the geocentric view between 1966 and 1985.

Planetariums

The geocentric (Ptolemaic) model of the solar system is still of interest to planetarium makers, as, for technical reasons, a Ptolemaic-type motion for the planet light apparatus has some advantages over a Copernican-type motion. The celestial sphere, still used for teaching purposes and sometimes for navigation, is also based on a geocentric system which in effect ignores parallax. However this effect is negligible at the scale of accuracy that applies to a planetarium.

Heliocentrism

Heliocentrism is the astronomical model in which the Earth and planets revolve around the Sun at the center of the Solar System. Historically, Heliocentrism was opposed to geocentrism, which placed the Earth at the center. The notion that the Earth revolves around the Sun had been proposed as early as the 3rd century BC by Aristarchus of Samos, but at least in the medieval world, Aristarchus's Heliocentrism attracted little attention—possibly because of the loss of scientific works of the Hellenistic Era.

Andreas Cellarius's illustration of the Copernican system, from the *Harmonia Macrocosmica* (1708).

It was not until the 16th century that a geometric mathematical model of a heliocentric system was presented, by the Renaissance mathematician, astronomer, and Catholic cleric Nicolaus Copernicus, leading to the Copernican Revolution. In the following century, Johannes Kepler elaborated upon and expanded this model to include elliptical orbits, and Galileo Galilei presented supporting observations made using a telescope.

With the observations of William Herschel, Friedrich Bessel, and others, astronomers realized that the sun, although near the center of Earth's solar system, was not the center of the universe.

Ancient and Medieval Astronomy

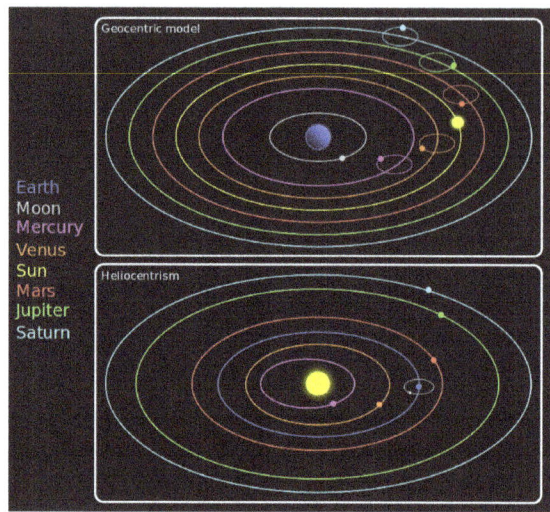

A hypothetical geocentric model of the solar system (upper panel) in comparison to the heliocentric model (lower panel).

While the sphericity of the Earth was widely recognized in Greco-Roman astronomy from at least the 3rd century BC, the Earth's daily rotation and yearly orbit around the Sun was never universally accepted until the Copernican Revolution.

While a moving Earth was proposed at least from the 4th century BC in Pythagoreanism, and a fully developed heliocentric model was developed by Aristarchus of Samos in the 3rd century BC, these ideas were not successful in replacing the view of a static spherical Earth, and from the 2nd century AD the predominant model, which would be inherited by medieval astronomy, was the geocentric model described in Ptolemy's *Almagest*.

The Ptolemaic system was a sophisticated astronomical system that managed to calculate the positions for the planets to a fair degree of accuracy. Ptolemy himself, in his *Almagest*, points out that any model for describing the motions of the planets is merely a mathematical device, and since there is no actual way to know which is true, the simplest model that gets the right numbers should be used. However, he rejected the idea of a spinning earth as absurd as he believed it would create huge winds. His planetary hypotheses were sufficiently real that the distances of moon, sun, planets and stars could be determined by treating orbits' celestial spheres as contiguous realities. This made the stars' distance less than 20 Astronomical Units, a regression, since Aristarchus of Samos's heliocentric scheme had centuries earlier necessarily placed the stars at least two orders of magnitude more distant.

Problems with Ptolemy's system were well recognized in medieval astronomy, and an increasing effort to criticize and improve it in the late medieval period eventually led to the Copernican heliocentrism developed in Renaissance astronomy.

Classical Antiquity

Pythagoreans

The non-geocentric model of the Universe was proposed by the Pythagorean philosopher Philolaus (d. 390 BC), who taught that at the center of the Universe was a "central fire", around which the Earth, Sun, Moon and Planets revolved in uniform circular motion. This system postulated the existence of a counter-earth collinear with the Earth and central fire, with the same period of revolution around the central fire as the Earth. The Sun revolved around the central fire once a year, and the stars were stationary. The Earth maintained the same hidden face towards the central fire, rendering both it and the "counter-earth" invisible from Earth. The Pythagorean concept of uniform circular motion remained unchallenged for approximately the next 2000 years, and it was to the Pythagoreans that Copernicus referred to show that the notion of a moving Earth was neither new nor revolutionary. Kepler gave an alternative explanation of the Pythagoreans' "central fire" as the Sun, "as most sects purposely hid[e] their teachings".

Heraclides of Pontus (4th century BC) said that the rotation of the Earth explained the apparent daily motion of the celestial sphere. It used to be thought that he believed Mercury and Venus to revolve around the Sun, which in turn (along with the other planets) revolves around the Earth. Macrobius Ambrosius Theodosius (AD 395–423) later described this as the "Egyptian System," stating that "it did not escape the skill of the Egyptians," though there is no other evidence it was known in ancient Egypt.

Aristarchus of Samos

The first person known to have proposed a heliocentric system, however, was Aristarchus of Samos (c. 270 BC). Like Eratosthenes, Aristarchus calculated the size of the Earth, and measured the size and distance of the Moon and Sun, in a treatise which has survived. From his estimates, he concluded that the Sun was six to seven times wider than the Earth and thus hundreds of times more voluminous. His writings on the heliocentric system are lost, but some information is known from surviving descriptions and critical commentary by his contemporaries, such as Archimedes. Some have suggested that his calculation of the relative size of the Earth and Sun led Aristarchus to conclude that it made more sense for the Earth to be moving than for the huge Sun to be moving around it. Though the original text has been lost, a reference in Archimedes' book *The Sand Reckoner* describes another work by Aristarchus in which he advanced an alternative hypothesis of the heliocentric model. Archimedes wrote:

You King Gelon are aware the 'universe' is the name given by most astronomers to the sphere the center of which is the center of the Earth, while its radius is equal to the straight line between the center of the Sun and the center of the Earth. This is the common account as you have heard from astronomers. But Aristarchus has brought out a book consisting of certain hypotheses, wherein it appears, as a consequence of the assumptions made, that the universe is many times greater than the 'universe' just mentioned. His hypotheses are that the fixed stars and the Sun remain unmoved, that the Earth revolves about the Sun on the circumference of a circle, the Sun lying in the middle of the orbit, and that the sphere of fixed stars, situated about the same center as the Sun, is so great that the circle in which he supposes the Earth to revolve bears such a proportion to the distance of the fixed stars as the center of the sphere bears to its surface.

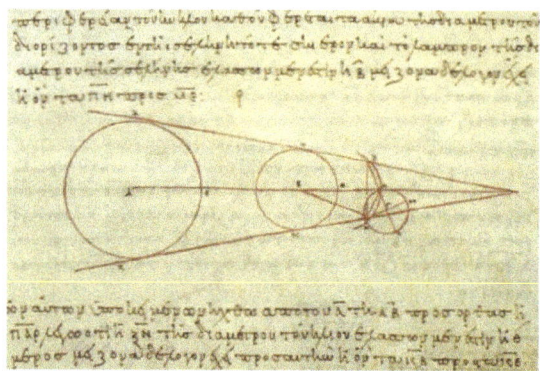

Aristarchus's 3rd century BC calculations on the relative sizes of the Earth, Sun and Moon, from a 10th-century CE Greek copy

Aristarchus believed the stars to be very far away, and saw this as the reason why there was no visible parallax, that is, an observed movement of the stars relative to each other as the Earth moved around the Sun. The stars are in fact much farther away than the distance that was generally assumed in ancient times, which is why stellar parallax is only detectable with telescopes.

Archimedes says that Aristarchus made the stars' distance larger, suggesting that he was answering the natural objection that Heliocentrism requires stellar parallactic oscillations. He apparently agreed to the point but placed the stars so distant as to make the parallactic motion invisibly minuscule. Thus Heliocentrism opened the way for realization that the universe was larger than the geocentrists taught.

Heliocentrism had been in conflict with religion before Copernicus: One of the few pieces of information we have about the reception of Aristarchus's heliocentric system comes from a passage in Plutarch's dialogue, *Concerning the Face which Appears in the Orb of the Moon*. According to one of Plutarch's characters in the dialogue, the philosopher Cleanthes had held that Aristarchus should be charged with impiety for "moving the hearth of the world".

Seleucus of Seleucia

Since Plutarch mentions the "followers of Aristarchus" in passing, it is likely that there were other astronomers in the Classical period who also espoused Heliocentrism, but whose work was lost. The only other astronomer from antiquity known by name who is known to have supported Aristarchus' heliocentric model was Seleucus of Seleucia (b. 190 BC), a Hellenistic astronomer who flourished a century after Aristarchus in the Seleucid empire. Seleucus adopted the heliocentric system of Aristarchus and is said to have proved the heliocentric theory. According to Bartel Leendert van der Waerden, Seleucus may have proved the heliocentric theory by determining the constants of a geometric model for the heliocentric theory and by developing methods to compute planetary positions using this model. He may have used early trigonometric methods that were available in his time, as he was a contemporary of Hipparchus. A fragment of a work by Seleucus has survived in Arabic translation, which was referred to by Rhazes (b. 865).

Alternatively, his explanation may have involved the phenomenon of tides, which he supposedly theorized to be caused by the attraction to the Moon and by the revolution of the Earth around the Earth-Moon 'center of mass'.

Late Antiquity

Nicholas of Cusa, 15th century, asked whether there was any reason to assert that any point was the center of the universe.

There were occasional speculations about Heliocentrism in Europe before Copernicus. In Roman Carthage, the pagan Martianus Capella (5th century A.D.) expressed the opinion that the planets Venus and Mercury did not go about the Earth but instead circled the Sun. Capella's model was discussed in the Early Middle Ages by various anonymous 9th-century commentators and Copernicus mentions him as an influence on his own work.

The Ptolemaic system was also received in Indian astronomy. Aryabhata (476–550), in his magnum opus *Aryabhatiya* (499), propounded a planetary model in which the Earth was taken to be spinning on its axis and the periods of the planets were given with respect to the Sun. He accurately calculated many astronomical constants, such as the periods of the planets, times of the solar and lunar eclipses, and the instantaneous motion of the Moon. Early followers of Aryabhata's model included Varahamihira, Brahmagupta, and Bhaskara II.

Medieval Islamic World

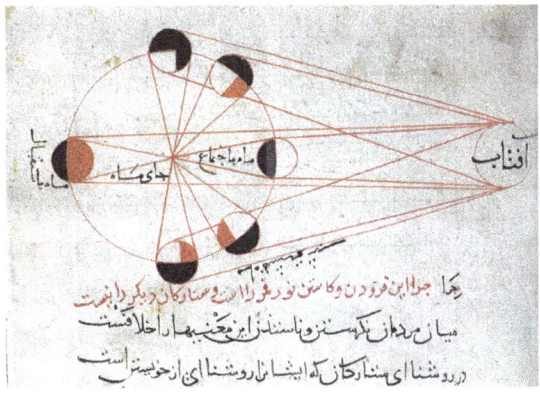

An illustration from al-Biruni's astronomical works, explains the different phases of the moon, with respect to the position of the sun. Al-Biruni suggested that if the Earth rotated on its axis this would be consistent with astronomical theory. He discussed Heliocentrism but considered it a problem of natural philosophy.

Muslim astronomers generally accepted the Ptolemaic system and the geocentric model.

Beginning in the 11th century, a tradition criticizing Ptolemy developed within Islamic astronomy, beginning with Ibn al-Haytham's *Al-Shukūk ʻalā Baṭlamyūs* ("Doubts Concerning Ptolemy"). Several Muslim scholars questioned the Earth's apparent immobility and centrality within the universe.

Abu Sa'id al-Sijzi (d. c. 1020) accepted that the Earth rotates around its axis.

According to Al-Biruni, Sijzi invented an astrolabe called *al-zūraqī* based on a belief held by some of his contemporaries "That the motion we see is due to the Earth's movement and not to that of the sky." The prevalence of this view is further confirmed by a reference from the 13th century which states:

According to the Geometers [or engineers] (*muhandisīn*), the earth is in constant circular motion, and what appears to be the motion of the heavens is actually due to the motion of the earth and not the stars.

Early in the 11th century Alhazen wrote a scathing critique of Ptolemy's model in his *Doubts on Ptolemy* (c. 1028), which some have interpreted to imply he was criticizing Ptolemy's geocentrism, but most agree that he was actually criticizing the details of Ptolemy's model rather than his geocentrism. Abu Rayhan Biruni (b. 973) discussed the possibility of whether the Earth rotated about its own axis and around the Sun, but in his *Masudic Canon*, he set forth the principles that the Earth is at the center of the universe and that it has no motion of its own. He was aware that if the Earth rotated on its axis, this would be consistent with his astronomical parameters, but he considered it a problem of natural philosophy rather than mathematics.

In the 12th century, some Islamic astronomers developed complete alternatives to the Ptolemaic system (although not heliocentric), such as Nur ad-Din al-Bitruji, who considered the Ptolemaic model as mathematical, and not physical. Al-Bitruji's alternative system spread through most of Europe in the 13th century, with debates and refutations of his ideas continued up to the 16th century.

Later Medieval Period

The Maragha school of astronomy in Ilkhanid-era Persia further developed "non-Ptolemaic" planetary models involving Earth's rotation. Notable astronomers of this school are Al-Urdi (d. 1266) Al-Katibi (d. 1277), and Al-Tusi (d. 1274).

The arguments and evidence used resemble those used by Copernicus to support the Earth's motion. The criticism of Ptolemy as developed by Averroes and by the Maragha school explicitly address the Earth's rotation but it did not arrive at explicit heliocentrism. The observations of the Maragha school were further improved at the Timurid-era Samarkand observatory under Qushji (1403–1474).

European scholarship in the later medieval period actively received astronomical models developed in the Islamic world and by the 13th century was well aware of the problems of the Ptolemaic model. In the 14th century, bishop Nicole Oresme discussed the possibility that the Earth rotated on its axis, while Cardinal Nicholas of Cusa in his *Learned Ignorance* asked whether there was any reason to assert that the Sun (or any other point) was the center of the universe. In parallel to a mystical definition of God, Cusa wrote that "Thus the fabric of the world (*machina mundi*) will *quasi* have its center everywhere and circumference nowhere."

In India, Nilakantha Somayaji (1444–1544), in his *Aryabhatiyabhasya*, a commentary on Aryabhata's *Aryabhatiya*, developed a computational system for a partially heliocentric planetary model, in which the planets orbit the Sun, which in turn orbits the Earth, similar to the Tychonic system later proposed by Tycho Brahe in the late 16th century. In the *Tantrasangraha* (1500), he further revised his planetary system, which was mathematically more accurate at predicting the heliocentric orbits of the interior planets than both the Tychonic and Copernican models, but like Indian astronomy in general fell short of proposing models of the universe. Nilakantha's planetary system also incorporated the Earth's rotation on its axis. Most astronomers of the Kerala school of astronomy and mathematics seem to have accepted his planetary model.

Renaissance-era Astronomy

European Astronomy Before Copernicus

Some historians maintain that the thought of the Maragha school, in particular the mathematical devices known as the Urdi lemma and the Tusi couple, influenced Renaissance-era European astronomy, and thus was indirectly received by Renaissance-era European astronomy and thus by Copernicus. Copernicus used such devices in the same planetary models as found in Arabic sources. Furthermore, the exact replacement of the equant by two epicycles used by Copernicus in the *Commentariolus* was found in an earlier work by Ibn al-Shatir (d. c. 1375) of Damascus. Ibn al-Shatir's lunar and Mercury models are also identical to those of Copernicus.

The state of knowledge on planetary theory received by Copernicus is summarized in Georg von Peuerbach's *Theoricae Novae Planetarum* (printed in 1472 by Regiomontanus). By 1470, the accuracy of observations by the Vienna school of astronomy, of which Peuerbach and Regiomontanus were members, was high enough to make the eventual development of heliocentrism inevitable, and indeed it is possible that Regiomontanus did arrive at an explicit theory of heliocentrism before his death in 1476, some 30 years before Copernicus. While the influence of the criticism of Ptolemy by Averroes on Renaissance thought is clear and explicit, the claim of direct influence of the Maragha school, postulated by Otto E. Neugebauer in 1957, remains an open question. Copernicus explicitly references several astronomers of the "Islamic Golden Age" (10th to 12th centuries) in *De Revolutionibus*: Albategnius (Al-Battani), Averroes (Ibn Rushd), Thebit (Thabit Ibn Qurra), Arzachel (Al-Zarqali), and Alpetragius (Al-Bitruji), but he does not show awareness of the existence of any of the later astronomers of the Maragha school.

It has been argued that Copernicus could have independently discovered the Tusi couple or took the idea from Proclus's *Commentary on the First Book of Euclid*, which Copernicus cited. Another possible source for Copernicus's knowledge of this mathematical device is the *Questiones de Spera* of Nicole Oresme, who described how a reciprocating linear motion of a celestial body could be produced by a combination of circular motions similar to those proposed by al-Tusi.

Copernican Heliocentrism

Portrait of Nicolaus Copernicus (1578)

Nicolaus Copernicus in his *De revolutionibus orbium coelestium* ("On the revolution of heavenly spheres", first printed in 1543 in Nuremberg), presented a discussion of a heliocentric model of the universe in much the same way as Ptolemy in the 2nd century had presented his geocentric model in his *Almagest*. Copernicus discussed the philosophical implications of his proposed system, elaborated it in geometrical detail, used selected astronomical observations to derive the parameters of his model, and wrote astronomical tables which enabled one to compute the past and future positions of the stars and planets. In doing so, Copernicus moved Heliocentrism from philosophical speculation to predictive geometrical astronomy. In reality, Copernicus's system did not predict the planets' positions any better than the Ptolemaic system. This theory resolved the issue of planetary retrograde motion by arguing that such motion was only perceived and apparent, rather than real: it was a parallax effect, as an object that one is passing seems to move backwards against the horizon. This issue was also resolved in the geocentric Tychonic system; the latter, however, while eliminating the major epicycles, retained as a physical reality the irregular back-and-forth motion of the planets, which Kepler characterized as a "pretzel".

Copernicus cited Aristarchus in an early (unpublished) manuscript of *De Revolutionibus* (which still survives), stating: "Philolaus believed in the mobility of the earth, and some even say that Aristarchus of Samos was of that opinion." However, in the published version he restricts himself to noting that in works by Cicero he had found an account of the theories of Hicetas and that Plutarch had provided him with an account of the Pythagoreans Heraclides Ponticus, Philolaus, and Ecphantus. These authors had proposed a moving earth, which did not, however, revolve around a central sun.

Reception in Early Modern Europe

Circulation of Commentariolus (Before 1515)

The first information about the heliocentric views of Nicolaus Copernicus was circulated in manuscript completed some time before May 1, 1514. Although only in manuscript, Copernicus' ideas were well known among astronomers and others. His ideas contradicted the then-prevailing understanding of the Bible. In the King James Bible (first published in 1611), First Chronicles 16:30 states that "the world also shall be stable, that it be not moved." Psalm 104:5 says, "[the Lord] Who laid the foundations of the earth, that it should not be removed for ever." Ecclesiastes 1:5 states that "The sun also ariseth, and the sun goeth down, and hasteth to his place where he arose."

Nonetheless, in 1533, Johann Albrecht Widmannstetter delivered in Rome a series of lectures outlining Copernicus' theory. The lectures were heard with interest by Pope Clement VII and several Catholic cardinals. On November 1, 1536, Archbishop of Capua Nikolaus von Schönberg wrote a letter to Copernicus from Rome encouraging him to publish a full version of his theory.

However, in 1539, Martin Luther said:

"There is talk of a new astrologer who wants to prove that the earth moves and goes around instead of the sky, the sun, the moon, just as if somebody were moving in a carriage or ship might hold that he was sitting still and at rest while the earth and the trees walked and moved. But that is how things are nowadays: when a man wishes to be clever he must . . . invent something special, and the way he does it must needs be the best! The fool wants to turn the whole art of astronomy upside-down. However, as Holy Scripture tells us, so did Joshua bid the sun to stand still and not the earth."

This was reported in the context of a conversation at the dinner table and not a formal statement of faith. Melanchthon, however, opposed the doctrine over a period of years.

De Revolutionibus

Nicolaus Copernicus published the definitive statement of his system in *De Revolutionibus* in 1543. Copernicus began to write it in 1506 and finished it in 1530, but did not publish it until the year of his death. Although he was in good standing with the Church and had dedicated the book to Pope Paul III, the published form contained an unsigned preface by Osiander defending the system and arguing that it was useful for computation even if its hypotheses were not necessarily true. Possibly because of that preface, the work of Copernicus inspired very little debate on whether it might be heretical during the next 60 years. There was an early suggestion among Dominicans that the teaching of Heliocentrism should be banned, but nothing came of it at the time.

Some years after the publication of *De Revolutionibus* John Calvin preached a sermon in which he denounced those who "pervert the order of nature" by saying that "the sun does not move and that it is the earth that revolves and that it turns".

On the other hand, Calvin is *not* responsible for another famous quotation which has often been misattributed to him: "Who will venture to place the authority of Copernicus above that of the Holy Spirit?" It has long been established that this line cannot be found in any of Calvin's works. It has been suggested that the quotation was originally sourced from the works of Lutheran theologian Abraham Calovius.

Tycho Brahe's Geo-heliocentric System c. 1587

Prior to the publication of *De Revolutionibus*, the most widely accepted system had been proposed by Ptolemy, in which the Earth was the center of the universe and all celestial bodies orbited it. Tycho Brahe, arguably the most accomplished astronomer of his time, advocated against Copernicus's heliocentric system and for an alternative to the Ptolemaic geocentric system: a geo-heliocentric system now known as the Tychonic system in which the five then known planets orbit the sun, while the sun and the moon orbit the earth.

Tycho appreciated the Copernican system, but objected to the idea of a moving Earth on the basis of physics, astronomy, and religion. The Aristotelian physics of the time (modern Newtonian physics was still a century away) offered no physical explanation for the motion of a massive body like Earth, whereas it could easily explain the motion of heavenly bodies by postulating that they were made of a different sort substance called aether that moved naturally. So Tycho said that the Copernican system "... expertly and completely circumvents all that is superfluous or discordant in the system of Ptolemy. On no point does it offend the principle of mathematics. Yet it ascribes to the Earth, that hulking, lazy body, unfit for motion, a motion as quick as that of the aethereal torches, and a triple motion at that." Likewise, Tycho took issue with the vast distances to the stars that Aristarchus and Copernicus had assumed in order to explain the lack of any visible parallax. Tycho had measured the apparent sizes of stars (now known to be illusory), and used geometry to calculate that in order to both have those apparent sizes and be as far away as Heliocentrism required, stars would have to be huge (much larger than the sun; the size of Earth's orbit or larger). Regarding this Tycho wrote, "Deduce these things geometrically if you like, and you will see how many absurdities (not to mention others) accompany this assumption [of the motion of the earth] by inference." He also cited the Copernican system's "opposition to the authority of Sacred Scripture in more than one place" as a reason why one might wish to reject it, and observed that his own geoheliocentric alternative "offended neither the principles of physics nor Holy Scripture".

The Jesuit astronomers in Rome were at first unreceptive to Tycho's system; the most prominent, Clavius, commented that Tycho was "confusing all of astronomy, because he wants to have Mars lower than the Sun." However, after the advent of the telescope showed problems with some geocentric models (by demonstrating that Venus circles the sun, for example), the Tychonic system and variations on that system became very popular among geocentrists, and the Jesuit astronomer Giovanni Battista Riccioli would continue Tycho's use of physics, stellar astronomy (now with a telescope), and religion to argue against Heliocentrism and for Tycho's system well into the seventeenth century.

Galileo Galilei

Starry Messenger

In the 17th century AD Galileo Galilei opposed the Roman Catholic Church by his strong support for Heliocentrism

Galileo was able to look at the night sky with the newly invented telescope. Then he published his discoveries in Sidereus Nuncius including (among other things) the moons of Jupiter and that

Venus exhibited a full range of phases. These discoveries were not consistent with the Ptolemeic model of the solar system. As the Jesuit astronomers confirmed Galileo's observations, the Jesuits moved toward Tycho's teachings.

In a Letter to the Grand Duchess Christina, Galileo defended Heliocentrism, and claimed it was not contrary to Scriptures. He took Augustine's position on Scripture: not to take every passage literally when the scripture in question is in a Bible book of poetry and songs, not a book of instructions or history. The writers of the Scripture wrote from the perspective of the terrestrial world, and from that vantage point the sun does rise and set. In fact, it is the Earth's rotation which gives the impression of the sun in motion across the sky.

1616 Ban Against Copernicanism

In February 1615, prominent Dominicans including Thomaso Caccini and Niccolò Lorini brought Galileo's writings on Heliocentrism to the attention of the Inquisition, because they appeared to violate Holy Scripture and the decrees of the Council of Trent. Cardinal and Inquisitor Robert Bellarmine was called upon to adjudicate, and wrote in April that treating Heliocentrism as a real phenomenon would be "a very dangerous thing," irritating philosophers and theologians, and harming "the Holy Faith by rendering Holy Scripture as false."

In January 1616 Msgr. Francesco Ingoli addressed an essay to Galileo disputing the Copernican system. Galileo later stated that he believed this essay to have been instrumental in the ban against Copernicanism that followed in February. According to Maurice Finocchiaro, Ingoli had probably been commissioned by the Inquisition to write an expert opinion on the controversy, and the essay provided the "chief direct basis" for the ban. The essay focused on eighteen physical and mathematical arguments against Heliocentrism. It borrowed primarily from the arguments of Tycho Brahe, and it notedly mentioned the problem that Heliocentrism requires the stars to be much larger than the sun. Ingoli wrote that the great distance to the stars in the heliocentric theory "clearly proves ... the fixed stars to be of such size, as they may surpass or equal the size of the orbit circle of the Earth itself." Ingoli included four theological arguments in the essay, but suggested to Galileo that he focus on the physical and mathematical arguments. Galileo did not write a response to Ingoli until 1624.

In February 1616, the Inquisition assembled a committee of theologians, known as qualifiers, who delivered their unanimous report condemning Heliocentrism as "foolish and absurd in philosophy, and formally heretical since it explicitly contradicts in many places the sense of Holy Scripture." The Inquisition also determined that the Earth's motion "receives the same judgement in philosophy and ... in regard to theological truth it is at least erroneous in faith." Bellarmine personally ordered Galileo

"to abstain completely from teaching or defending this doctrine and opinion or from discussing it... to abandon completely... the opinion that the sun stands still at the center of the world and the earth moves, and henceforth not to hold, teach, or defend it in any way whatever, either orally or in writing."

—Bellarmine and the Inquisition's injunction against Galileo, 1616

In March, after the Inquisition's injunction against Galileo, the papal Master of the Sacred Palace,

Congregation of the Index, and Pope banned all books and letters advocating the Copernican system, which they called "the false Pythagorean doctrine, altogether contrary to Holy Scripture." In 1618 the Holy Office recommended that a modified version of Copernicus' *De Revolutionibus* be allowed for use in calendric calculations, though the original publication remained forbidden until 1758.

Epitome Astronomia Copernicanae

In *Astronomia nova* (1609), Johannes Kepler had used an elliptical orbit to explain the motion of Mars. In *Epitome astronomiae Copernicanae* he developed a heliocentric model of the solar system in which all the planets have elliptical orbits. This provided significantly increased accuracy in predicting the position of the planets. Kepler's ideas were not immediately accepted. Galileo for example completely ignored Kepler's work. Kepler proposed Heliocentrism as a physical description of the solar system and *Epitome astronomia Copernicanae* was placed on the index of prohibited books despite Kepler being a Protestant.

Dialogue Concerning the Two Chief World Systems

Pope Urban VIII encouraged Galileo to publish the pros and cons of Heliocentrism. Galileo's response, *Dialogue concerning the two chief world systems,* clearly advocated Heliocentrism, despite his declaration in the preface that

I will endeavour to show that all experiments that can be made upon the Earth are insufficient means to conclude for its mobility but are indifferently applicable to the Earth, movable or immovable...

and his straightforward statement,

I might very rationally put it in dispute, whether there be any such centre in nature, or no; being that neither you nor any one else hath ever proved, whether the World be finite and figurate, or else infinite and interminate; yet nevertheless granting you, for the present, that it is finite, and of a terminate Spherical Figure, and that thereupon it hath its centre...

Some ecclesiastics also interpreted the book as characterizing the Pope as a simpleton, since his viewpoint in the dialogue was advocated by the character Simplicio. Urban VIII became hostile to Galileo and he was again summoned to Rome. Galileo's trial in 1633 involved making fine distinctions between "teaching" and "holding and defending as true". For advancing heliocentric theory Galileo was forced to recant Copernicanism and was put under house arrest for the last few years of his life.

According to J. L. Heilbron, informed contemporaries of Galileo's:

"appreciated that the reference to heresy in connection with Galileo or Copernicus had no general or theological significance."

Age of Reason

René Descartes postponed, and ultimately never finished, his treatise *The World*, which included a heliocentric model, but the Galileo affair did little to slow the spread of Heliocentrism across Europe, as Kepler's *Epitome of Copernican Astronomy* became increasingly influential in the coming decades. By 1686 the model was well enough established that the general public was reading about it in *Conversations on the Plurality of Worlds*, published in France by Bernard le Bovier de Fontenelle and translated into English and other languages in the coming years. It has been called "one of the first great popularizations of science."

In 1687, Isaac Newton published *Philosophiæ Naturalis Principia Mathematica*, which provided an explanation for Kepler's laws in terms of universal gravitation and what came to be known as Newton's laws of motion. This placed Heliocentrism on a firm theoretical foundation, although Newton's Heliocentrism was of a somewhat modern kind. Already in the mid-1680s he recognized the "deviation of the Sun" from the centre of gravity of the solar system. For Newton it was not precisely the centre of the Sun or any other body that could be considered at rest, but "the common centre of gravity of the Earth, the Sun and all the Planets is to be esteem'd the Centre of the World", and this centre of gravity "either is at rest or moves uniformly forward in a right line". Newton adopted the "at rest" alternative in view of common consent that the centre, wherever it was, was at rest.

Meanwhile, the Church remained opposed to Heliocentrism as a literal description, but this did not by any means imply opposition to all astronomy; indeed, it needed observational data to maintain its calendar. In support of this effort it allowed the cathedrals themselves to be used as solar observatories called *meridiane*; i.e., they were turned into "reverse sundials", or gigantic pinhole cameras, where the Sun's image was projected from a hole in a window in the cathedral's lantern onto a meridian line.

In 1664, Pope Alexander VII published his *Index Librorum Prohibitorum Alexandri VII Pontificis Maximi jussu editus* (Index of Prohibited Books, published by order of Alexander VII, P.M.) which included all previous condemnations of heliocentric books.

In the mid-eighteenth century the Church's opposition began to fade. An annotated copy of Newton's *Principia* was published in 1742 by Fathers le Seur and Jacquier of the Franciscan Minims, two Catholic mathematicians, with a preface stating that the author's work assumed Heliocentrism and could not be explained without the theory. In 1758 the Catholic Church dropped the general prohibition of books advocating Heliocentrism from the *Index of Forbidden Books*. The Observatory of the Roman College was established by Pope Clement XIV in 1774 (nationalized in 1878, but re-founded by Pope Leo XIII as the Vatican Observatory in 1891). In spite of dropping its active resistance to Heliocentrism, the Catholic Church did not lift the prohibition of uncensored versions of Copernicus's *De Revolutionibus* or Galileo's *Dialogue*. The affair was revived in 1820, when the Master of the Sacred Palace (the Church's chief censor), Filippo Anfossi, refused to license a book by a Catholic canon, Giuseppe Settele, because it openly treated heliocentrism as a physical fact. Settele appealed to pope Pius VII. After the matter had been reconsidered by the Congregation of the Index and the Holy Office, Anfossi's decision was overturned. Pius VII approved a decree in 1822 by the Sacred Congregation of the Inquisition to allow the printing of heliocentric books in Rome. Copernicus's *De Revolutionibus* and Galileo's *Dialogue* were then subsequently omitted from the next edition of the *Index* when it appeared in 1835.

Reception in Judaism

Already in the Talmud, Greek philosophy and science under general name "Greek wisdom" were considered dangerous. They were put under ban then and later for some periods.

The first Jewish scholar to describe the Copernican system, albeit without mentioning Copernicus by name, was Maharal of Prague, his book "Be'er ha-Golah" (1593). Maharal makes an argument of radical skepticism, arguing that no scientific theory can be reliable, which he illustrates by the new-fangled theory of heliocentrism upsetting even the most fundamental views on the cosmos.

Copernicus is mentioned in the books of David Gans (1541–1613), who worked with Tycho Brahe and Johannes Kepler. Gans wrote two books on astronomy in Hebrew: a short one "Magen David" (1612) and a full one "Nehmad veNaim" (published only in 1743). He described objectively three systems: Ptolemy, Copernicus and of Tycho Brahe without taking sides. Joseph Solomon Delmedigo (1591–1655) in his "Elim" (1629) says that the arguments of Copernicus are so strong, that only an imbecile will not accept them. Delmedigo studied at Padua and was acquainted with Galileo.

An actual controversy on the Copernican model within Judaism arises only in the early 18th century. Most authors in this period accept Copernican heliocentrism, with opposition from David Nieto and Tobias Cohn. Both of these authors argued against heliocentrism on grounds of contradictions to scripture. Nieto merely rejected the new system on those grounds without much passion, whereas Cohn went so far as to call Copernicus "a first-born of Satan", though he also acknowledged that he would have found it difficult to counter one particular objection based on a passage from the Talmud. In the later 18th and 19th centuries, there were no explicit attacks on Heliocentrism, although some Rabbis expressed a position of agnosticism.

In the 20th century, R. M.M. Schneerson argued that the question of heliocentrism vs. geocentrism is obsolete because of the relativity of motion.

The View of Modern Science

Kepler's laws of planetary motion were used as arguments in favor of the heliocentric hypothesis. Three apparent proofs of the heliocentric hypothesis were provided in 1727 by James Bradley, in 1838 by Friedrich Wilhelm Bessel and in 1851 by Foucault. Bessel proved that the parallax of a star was greater than zero by measuring the parallax of 0.314 arcseconds of a star named 61 Cygni. In the same year Friedrich Georg Wilhelm Struve and Thomas Henderson measured the parallaxes of other stars, Vega and Alpha Centauri.

The thinking that the heliocentric view was also not true in a strict sense was achieved in steps. That the Sun was not the center of the universe, but one of innumerable stars, was strongly advocated by the mystic Giordano Bruno. Over the course of the 18th and 19th centuries, the status of the Sun as merely one star among many became increasingly obvious. By the 20th century, even before the discovery that there are many galaxies, it was no longer an issue.

The concept of an absolute velocity, including being "at rest" as a particular case, is ruled out by the principle of relativity, also eliminating any obvious "center" of the universe as a natural origin of coordinates. Some forms of Mach's principle consider the frame at rest with respect to the distant masses in the universe to have special properties.

Even if the discussion is limited to the solar system, the Sun is not at the geometric center of any planet's orbit, but rather approximately at one focus of the elliptical orbit. Furthermore, to the extent that a planet's mass cannot be neglected in comparison to the Sun's mass, the center of gravity of the solar system is displaced slightly away from the center of the Sun. (The masses of the planets, mostly Jupiter, amount to 0.14% of that of the Sun.) Therefore, a hypothetical astronomer on an extrasolar planet would observe a small "wobble" in the Sun's motion.

Modern use of *Geocentric* and *Heliocentric*

In modern calculations the terms "geocentric" and "heliocentric" are often used to refer to reference frames. In such systems the origin in the center of mass of the Earth, of the Earth–Moon system, of the Sun, of the Sun plus the major planets, or of the entire solar system can be selected; see center-of-mass frame. Right Ascension and Declination are examples of geocentric coordinates, used in Earth-based observations, while the heliocentric latitude and longitude are used for orbital calculations. This leads to such terms as "heliocentric velocity" and "heliocentric angular momentum". In this heliocentric picture, any planet of the Solar System can be used as a source of mechanical energy because it moves relatively to the Sun. A smaller body (either artificial or natural) may gain heliocentric velocity due to gravity assist – this effect can change the body's mechanical energy in heliocentric reference frame (although it will not changed in the planetary one). However, such selection of "geocentric" or "heliocentric" frames is merely a matter of computation. It does not have philosophical implications and does not constitute a distinct physical or scientific model. From the point of view of General Relativity, inertial reference frames do not exist at all, and any practical reference frame is only an approximation to the actual space-time, which can have higher or lower precision.

Big Bang

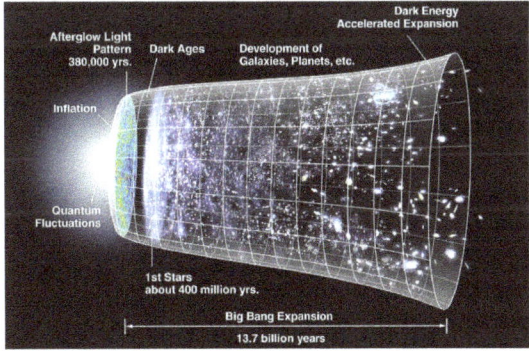

Timeline of the metric expansion of space, where space (including hypothetical non-observable portions of the universe) is represented at each time by the circular sections. On the left, the dramatic expansion occurs in the inflationary epoch; and at the center, the expansion accelerates (artist's concept; not to scale).

The Big Bang theory is the prevailing cosmological model for the universe from the earliest known periods through its subsequent large-scale evolution. The model describes how the universe expanded from a very high density and high temperature state, and offers a comprehensive explanation for a broad range of phenomena, including the abundance of light elements, the cosmic microwave background, large scale structure and Hubble's Law. If the known laws of physics are extrapolated to the highest density regime, the result is a singularity which is typically associated with the Big Bang. Detailed measurements of the expansion rate of the universe place this moment at approximately 13.8 billion years ago, which is thus considered the age of the universe. After the initial expansion, the universe cooled sufficiently to allow the formation of subatomic particles, and later simple atoms. Giant clouds of these primordial elements later coalesced through gravity in halos of dark matter, eventually forming the stars and galaxies visible today.

Since Georges Lemaître first noted in 1927 that an expanding universe could be traced back in time to an originating single point, scientists have built on his idea of cosmic expansion. While the scientific community was once divided between supporters of two different expanding universe theories, the Big Bang and the Steady State theory, empirical evidence provides strong support for the former. In 1929, from analysis of galactic redshifts, Edwin Hubble concluded, that galaxies are drifting apart; this is important observational evidence consistent with the hypothesis of an expanding universe. In 1964, the cosmic microwave background radiation was discovered, which was crucial evidence in favor of the Big Bang model, since that theory predicted the existence of background radiation throughout the universe before it was discovered. More recently, measurements of the redshifts of supernovae indicate that the expansion of the universe is accelerating, an observation attributed to dark energy's existence. The known physical laws of nature can be used to calculate the characteristics of the universe in detail back in time to an initial state of extreme density and temperature.

Overview

American astronomer Edwin Hubble observed that the distances to faraway galaxies were strongly correlated with their redshifts. This was interpreted to mean that all distant galaxies and clusters are receding away from our vantage point with an apparent velocity proportional to their distance: that is, the farther they are, the faster they move away from us, regardless of direction. Assuming the Copernican principle (that the Earth is not the center of the universe), the only remaining interpretation is that all observable regions of the universe are receding from all others. Since we know that the distance between galaxies increases today, it must mean that in the past galaxies were closer together. The continuous expansion of the universe implies that the universe was denser and hotter in the past.

Large particle accelerators can replicate the conditions that prevailed after the early moments of the universe, resulting in confirmation and refinement of the details of the Big Bang model. However, these accelerators can only probe so far into high energy regimes. Consequently, the state of the universe in the earliest instants of the Big Bang expansion is still poorly understood and an area of open investigation and speculation.

The first subatomic particles to be formed included protons, neutrons, and electrons. Though simple atomic nuclei formed within the first three minutes after the Big Bang, thousands of years passed before the first electrically neutral atoms formed. The majority of atoms produced by the Big Bang were hydrogen, along with helium and traces of lithium. Giant clouds of these primordial elements later coalesced through gravity to form stars and galaxies, and the heavier elements were synthesized either within stars or during supernovae.

The Big Bang theory offers a comprehensive explanation for a broad range of observed phenomena, including the abundance of light elements, the cosmic microwave background, large scale structure, and Hubble's Law. The framework for the Big Bang model relies on Albert Einstein's theory of general relativity and on simplifying assumptions such as homogeneity and isotropy of space. The governing equations were formulated by Alexander Friedmann, and similar solutions were worked on by Willem de Sitter. Since then, astrophysicists have incorporated observational and theoretical additions into the Big Bang model, and its parametrization as the Lambda-CDM model serves as the framework for current investigations of theoretical cosmology. The Lambda-CDM model is the current "standard model" of Big Bang cosmology, consensus is that it is the simplest model that can account for the various measurements and observations relevant to cosmology.

Timeline

Singularity

Extrapolation of the expansion of the universe backwards in time using general relativity yields an infinite density and temperature at a finite time in the past. This singularity indicates that general relativity is not an adequate description of the laws of physics in this regime. How closely models based on general relativity alone can be used to extrapolate toward the singularity is debated—certainly no closer than the end of the Planck epoch.

This primordial singularity is itself sometimes called "the Big Bang", but the term can also refer to a more generic early hot, dense phase of the universe. In either case, "the Big Bang" as an event is also colloquially referred to as the "birth" of our universe since it represents the point in history where the universe can be verified to have entered into a regime where the laws of physics as we understand them (specifically general relativity and the standard model of particle physics) work. Based on measurements of the expansion using Type Ia supernovae and measurements of temperature fluctuations in the cosmic microwave background, the time that has passed since that event — otherwise known as the "age of the universe" — is 13.799 ± 0.021 billion years. The agreement of independent measurements of this age supports the ΛCDM model that describes in detail the characteristics of the universe.

Inflation and Baryogenesis

The earliest phases of the Big Bang are subject to much speculation. In the most common models the universe was filled homogeneously and isotropically with a very high energy density and huge temperatures and pressures and was very rapidly expanding and cooling. Approximately 10^{-37} seconds into the expansion, a phase transition caused a cosmic inflation, during which the universe grew exponentially during which time density fluctuations that occurred because of the uncertainty principle were amplified into the seeds that would later form the large-scale structure of the universe. After inflation stopped, reheating occurred until the universe obtained the temperatures required for the production of a quark–gluon plasma as well as all other elementary particles. Temperatures were so high that the random motions of particles were at relativistic speeds, and particle–antiparticle pairs of all kinds were being continuously created and destroyed in collisions. At some point, an unknown reaction called baryogenesis violated the conservation of baryon number, leading to a very small excess of quarks and leptons over antiquarks and antileptons—of the order of one part in 30 million. This resulted in the predominance of matter over antimatter in the present universe.

Cooling

The universe continued to decrease in density and fall in temperature, hence the typical energy of each particle was decreasing. Symmetry breaking phase transitions put the fundamental forces of physics and the parameters of elementary particles into their present form. After about 10^{-11} seconds, the picture becomes less speculative, since particle energies drop to values that can be attained in particle accelerators. At about 10^{-6} seconds, quarks and gluons combined to form baryons such as protons and neutrons. The small excess of quarks over antiquarks led to a small excess of baryons over antibaryons. The temperature was now no longer high enough to create new proton–antiproton pairs (similarly for neutrons–antineutrons), so a mass annihilation immediately followed, leaving just one in 10^{10} of the original protons and neutrons, and none of their antiparticles. A similar process happened at about 1 second for electrons and positrons. After these annihilations, the remaining protons, neutrons and electrons were no longer moving relativistically and the energy density of the universe was dominated by photons (with a minor contribution from neutrinos).

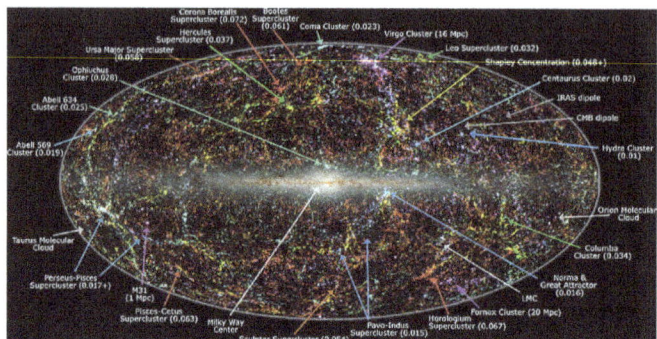

Panoramic view of the entire near-infrared sky reveals the distribution of galaxies beyond the Milky Way. Galaxies are color-coded by redshift.

A few minutes into the expansion, when the temperature was about a billion (one thousand million) kelvin and the density was about that of air, neutrons combined with protons to form the universe's deuterium and helium nuclei in a process called Big Bang nucleosynthesis. Most protons remained uncombined as hydrogen nuclei.

As the universe cooled, the rest mass energy density of matter came to gravitationally dominate that of the photon radiation. After about 379,000 years, the electrons and nuclei combined into atoms (mostly hydrogen); hence the radiation decoupled from matter and continued through space largely unimpeded. This relic radiation is known as the cosmic microwave background radiation. The chemistry of life may have begun shortly after the Big Bang, 13.8 billion years ago, during a habitable epoch when the universe was only 10–17 million years old.

Structure Formation

Over a long period of time, the slightly denser regions of the nearly uniformly distributed matter gravitationally attracted nearby matter and thus grew even denser, forming gas clouds, stars, galaxies, and the other astronomical structures observable today. The details of this process depend on the amount and type of matter in the universe. The four possible types of matter are known as cold dark matter, warm dark matter, hot dark matter, and baryonic matter. The best measurements available (from WMAP) show that the data is well-fit by a Lambda-CDM model in which dark matter is assumed to be cold (warm dark matter is ruled out by early reionization), and is estimated to make up about 23% of the matter/energy of the universe, while baryonic matter makes up about 4.6%. In an "extended model" which includes hot dark matter in the form of neutrinos, then if the "physical baryon density" $\Omega_b h^2$ is estimated at about 0.023 (this is different from the 'baryon density' Ω_b expressed as a fraction of the total matter/energy density, which as noted above is about 0.046), and the corresponding cold dark matter density $\Omega_c h^2$ is about 0.11, the corresponding neutrino density $\Omega_v h^2$ is estimated to be less than 0.0062.

Abell 2744 galaxy cluster - Hubble Frontier Fields view.

Cosmic Acceleration

Independent lines of evidence from Type Ia supernovae and the CMB imply that the universe today is dominated by a mysterious form of energy known as dark energy, which apparently permeates all of space. The observations suggest 73% of the total energy density of today's universe is in this form. When the universe was very young, it was likely infused with dark energy, but with less space and everything closer together, gravity predominated, and it was slowly braking the expansion. But eventually, after numerous billion years of expansion, the growing abundance of dark energy caused the expansion of the universe to slowly begin to accelerate.

Dark energy in its simplest formulation takes the form of the cosmological constant term in Einstein's field equations of general relativity, but its composition and mechanism are unknown and, more generally, the details of its equation of state and relationship with the Standard Model of particle physics continue to be investigated both through observation and theoretically.

All of this cosmic evolution after the inflationary epoch can be rigorously described and modeled by the ΛCDM model of cosmology, which uses the independent frameworks of quantum mechanics and Einstein's General Relativity. There is no well-supported model describing the action prior to 10^{-15} seconds or so. Apparently a new unified theory of quantum gravitation is needed to break this barrier. Understanding this earliest of eras in the history of the universe is currently one of the greatest unsolved problems in physics.

Features of the Model

The Big Bang theory depends on two major assumptions: the universality of physical laws and the cosmological principle. The cosmological principle states that on large scales the universe is homogeneous and isotropic.

These ideas were initially taken as postulates, but today there are efforts to test each of them. For example, the first assumption has been tested by observations showing that largest possible deviation of the fine structure constant over much of the age of the universe is of order 10^{-5}. Also, general relativity has passed stringent tests on the scale of the Solar System and binary stars.

If the large-scale universe appears isotropic as viewed from Earth, the cosmological principle can be derived from the simpler Copernican principle, which states that there is no preferred (or special) observer or vantage point. To this end, the cosmological principle has been confirmed to a level of 10^{-5} via observations of the CMB. The universe has been measured to be homogeneous on the largest scales at the 10% level.

Expansion of Space

General relativity describes spacetime by a metric, which determines the distances that separate nearby points. The points, which can be galaxies, stars, or other objects, are themselves specified using a coordinate chart or "grid" that is laid down over all spacetime. The cosmological principle implies, that the metric should be homogeneous and isotropic on large scales, which uniquely singles out the Friedmann–Lemaître–Robertson–Walker metric (FLRW metric). This metric contains a scale factor, which describes how the size of the universe changes with time. This enables a convenient choice of a coordinate system to be made, called comoving coordinates. In this coordinate system, the grid expands along with the universe, and objects that are moving only because of the expansion of the universe, remain at fixed points on the grid. While their *coordinate* distance (comoving distance) remains constant, the *physical* distance between two such co-moving points expands proportionally with the scale factor of the universe.

The Big Bang is not an explosion of matter moving outward to fill an empty universe. Instead, space itself expands with time everywhere and increases the physical distance between two co-moving points. In other words, the Big Bang is not an explosion *in space*, but rather an expansion *of space*. Because the FLRW metric assumes a uniform distribution of mass and energy, it applies to our universe only on large scales—local concentrations of matter such as our galaxy are gravitationally bound and as such do not experience the large-scale expansion of space.

Horizons

An important feature of the Big Bang spacetime is the presence of particle horizons. Since the universe has a finite age, and light travels at a finite speed, there may be events in the past, whose light has not had time to reach us. This places a limit or a *past horizon* on the most distant objects that can be observed. Conversely, because space is expanding, and more distant objects are receding ever more quickly, light emitted by us today may never "catch up" to very distant objects. This defines a *future horizon*, which limits the events in the future that we will be able to influence. The presence of either type of horizon depends on the details of the FLRW model that describes our universe.

Our understanding of the universe back to very early times suggests, that there is a past horizon, though in practice, our view is also limited by the opacity of the universe at early times. So our view cannot extend further backward in time, though the horizon recedes in space. If the expansion of the universe continues to accelerate, there is a future horizon as well.

History

Etymology

English astronomer Fred Hoyle is credited with coining the term "Big Bang" during a 1949 BBC radio broadcast. It is popularly reported that Hoyle, who favored an alternative "steady state" cosmological model, intended this to be pejorative, but Hoyle explicitly denied this and said it was just a striking image meant to highlight the difference between the two models.

Development

Hubble eXtreme Deep Field (XDF)

XDF size compared to the size of the Moon - several thousand galaxies, each consisting of billions of stars, are in this small view.

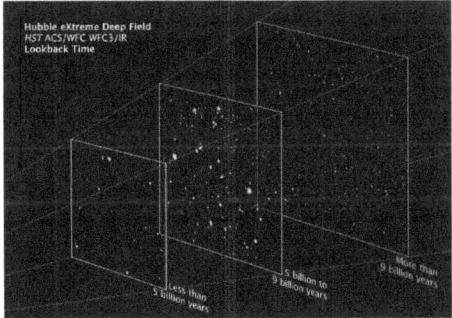

XDF image shows fully mature galaxies in the foreground plane - nearly mature galaxies from 5 to 9 billion years ago - protogalaxies, blazing with young stars, beyond 9 billion years.

The Big Bang theory developed from observations of the structure of the universe and from theoretical considerations. In 1912 Vesto Slipher measured the first Doppler shift of a "spiral nebula" (spiral nebula is the obsolete term for spiral galaxies), and soon discovered that almost all such nebulae were receding from Earth. He did not grasp the cosmological implications of this fact, and indeed at the time it was highly controversial whether or not these nebulae were "island universes" outside our Milky Way. Ten years later, Alexander Friedmann, a Russian cosmologist and mathematician, derived the Friedmann equations from Albert Einstein's equations of general relativity, showing that the universe might be expanding in contrast to the static universe model advocated by Einstein at that time. In 1924 Edwin Hubble's measurement of the great distance to the nearest spiral nebulae showed that these systems were indeed other galaxies. Independently deriving Friedmann's equations in 1927, Georges Lemaître, a Belgian physicist and Roman Catholic priest, proposed that the inferred recession of the nebulae was due to the expansion of the universe.

In 1931 Lemaître went further and suggested that the evident expansion of the universe, if projected back in time, meant that the further in the past the smaller the universe was, until at some finite time in the past all the mass of the universe was concentrated into a single point, a "primeval atom" where and when the fabric of time and space came into existence.

Starting in 1924, Hubble painstakingly developed a series of distance indicators, the forerunner of the cosmic distance ladder, using the 100-inch (2.5 m) Hooker telescope at Mount Wilson Observatory. This allowed him to estimate distances to galaxies whose redshifts had already been measured, mostly by Slipher. In 1929 Hubble discovered a correlation between distance and recession velocity—now known as Hubble's law. Lemaître had already shown that this was expected, given the cosmological principle.

In the 1920s and 1930s almost every major cosmologist preferred an eternal steady state universe, and several complained that the beginning of time implied by the Big Bang imported religious concepts into physics; this objection was later repeated by supporters of the steady state theory. This perception was enhanced by the fact that the originator of the Big Bang theory, Monsignor Georges Lemaître, was a Roman Catholic priest. Arthur Eddington agreed with Aristotle that the universe did not have a beginning in time, viz., that matter is eternal. A beginning in time was "repugnant" to him. Lemaître, however, thought that

If the world has begun with a single quantum, the notions of space and time would altogether fail to have any meaning at the beginning; they would only begin to have a sensible meaning when the original quantum had been divided into a sufficient number of quanta. If this suggestion is correct, the beginning of the world happened a little before the beginning of space and time.

During the 1930s other ideas were proposed as non-standard cosmologies to explain Hubble's observations, including the Milne model, the oscillatory universe (originally suggested by Friedmann, but advocated by Albert Einstein and Richard Tolman) and Fritz Zwicky's tired light hypothesis.

After World War II, two distinct possibilities emerged. One was Fred Hoyle's steady state model, whereby new matter would be created as the universe seemed to expand. In this model the universe is roughly the same at any point in time. The other was Lemaître's Big Bang theory, advocated and developed by George Gamow, who introduced big bang nucleosynthesis (BBN) and whose associates, Ralph Alpher and Robert Herman, predicted the cosmic microwave background radiation (CMB). Ironically, it was Hoyle who coined the phrase that came to be applied to Lemaître's theory, referring to it as "this *big bang* idea" during a BBC Radio broadcast in March 1949. For a while, support was split between these two theories. Eventually, the observational evidence, most notably from radio source counts, began to favor Big Bang over Steady State. The discovery and confirmation of the cosmic microwave background radiation in 1964 secured the Big Bang as the best theory of the origin and evolution of the universe. Much of the current work in cosmology includes understanding how galaxies form in the context of the Big Bang, understanding the physics of the universe at earlier and earlier times, and reconciling observations with the basic theory.

In 1968 and 1970 Roger Penrose, Stephen Hawking, and George F. R. Ellis published papers where they showed that mathematical singularities were an inevitable initial condition of general relativistic models of the Big Bang. Then, from the 1970s to the 1990s, cosmologists worked on characterizing the features of the Big Bang universe and resolving outstanding problems. In 1981, Alan Guth made a breakthrough in theoretical work on resolving certain outstanding theoretical problems in the Big Bang theory with the introduction of an epoch of rapid expansion in the early universe he called "inflation". Meanwhile, during these decades, two questions in observational cosmology that generated much discussion and disagreement were over the precise values of the Hubble Constant and the matter-density of the universe (before the discovery of dark energy, thought to be the key predictor for the eventual fate of the universe).

In the mid-1990s, observations of certain globular clusters appeared to indicate, that they were about 15 billion years old, which conflicted with most then-current estimates of the age of the universe (and indeed with the age measured today). This issue was later resolved when new computer simulations, which included the effects of mass loss due to stellar winds, indicated a much younger age for globular clusters. While there still remain some questions as to how accurately the ages of the clusters are measured, globular clusters are of interest to cosmology as some of the oldest objects in the universe.

Significant progress in Big Bang cosmology have been made since the late 1990s as a result of advances in telescope technology as well as the analysis of data from satellites such as COBE, the Hubble Space Telescope and WMAP. Cosmologists now have fairly precise and accurate measurements of many of the parameters of the Big Bang model, and have made the unexpected discovery that the expansion of the universe appears to be accelerating.

Observational Evidence

Artist's depiction of the WMAP satellite gathering data to help scientists understand the Big Bang

"[The] big bang picture is too firmly grounded in data from every area to be proved invalid in its general features."

Lawrence Krauss

The earliest and most direct observational evidence of the validity of the theory are the expansion of the universe according to Hubble's law (as indicated by the redshifts of galaxies), discovery and measurement of the cosmic microwave background and the relative abundances of light elements produced by Big Bang nucleosynthesis. More recent evidence includes observations of galaxy formation and evolution, and the distribution of large-scale cosmic structures, These are sometimes called the "four pillars" of the Big Bang theory.

Precise modern models of the Big Bang appeal to various exotic physical phenomena that have not been observed in terrestrial laboratory experiments or incorporated into the Standard Model of particle physics. Of these features, dark matter is currently subjected to the most active laboratory investigations. Remaining issues include the cuspy halo problem and the dwarf galaxy problem of cold dark matter. Dark energy is also an area of intense interest for scientists, but it is not clear whether direct detection of dark energy will be possible. Inflation and baryogenesis remain more speculative features of current Big Bang models. Viable, quantitative explanations for such phenomena are still being sought. These are currently unsolved problems in physics.

Hubble's Law and the Expansion of Space

Observations of distant galaxies and quasars show that these objects are redshifted—the light emitted from them has been shifted to longer wavelengths. This can be seen by taking a frequency spectrum of an object and matching the spectroscopic pattern of emission lines or absorption lines corresponding to atoms of the chemical elements interacting with the light. These redshifts are uniformly isotropic, distributed evenly among the observed objects in all directions. If the redshift is interpreted as a Doppler shift, the recessional velocity of the object can be calculated. For some galaxies, it is possible to estimate distances via the cosmic distance ladder. When the recessional velocities are plotted against these distances, a linear relationship known as Hubble's law is observed:

$$v = H_0 D,$$

where

- v is the recessional velocity of the galaxy or other distant object,

- D is the comoving distance to the object, and

- H_0 is Hubble's constant, measured to be $70.4^{+1.3}_{-1.4}$ km/s/Mpc by the WMAP probe.

Hubble's law has two possible explanations. Either we are at the center of an explosion of galaxies—which is untenable given the Copernican principle—or the universe is uniformly expanding everywhere. This universal expansion was predicted from general relativity by Alexander Friedmann in 1922 and Georges Lemaître in 1927, well before Hubble made his 1929 analysis and observations, and it remains the cornerstone of the Big Bang theory as developed by Friedmann, Lemaître, Robertson, and Walker.

The theory requires the relation $v = HD$ to hold at all times, where D is the comoving distance, v is the recessional velocity, and v, H, and D vary as the universe expands (hence we write H_0 to denote the present-day Hubble "constant"). For distances much smaller than the size of the observable universe, the Hubble redshift can be thought of as the Doppler shift corresponding to the recession velocity v. However, the redshift is not a true Doppler shift, but rather the result of the expansion of the universe between the time the light was emitted and the time that it was detected.

That space is undergoing metric expansion is shown by direct observational evidence of the Cosmological principle and the Copernican principle, which together with Hubble's law have no other explanation. Astronomical redshifts are extremely isotropic and homogeneous, supporting the Cosmological principle that the universe looks the same in all directions, along with much other evidence. If the redshifts were the result of an explosion from a center distant from us, they would not be so similar in different directions.

Measurements of the effects of the cosmic microwave background radiation on the dynamics of distant astrophysical systems in 2000 proved the Copernican principle, that, on a cosmological scale, the Earth is not in a central position. Radiation from the Big Bang was demonstrably warmer at earlier times throughout the universe. Uniform cooling of the cosmic microwave background over billions of years is explainable only if the universe is experiencing a metric expansion, and excludes the possibility that we are near the unique center of an explosion.

Cosmic Microwave Background Radiation

In 1964 Arno Penzias and Robert Wilson serendipitously discovered the cosmic background radiation, an omnidirectional signal in the microwave band. Their discovery provided substantial confirmation of the big-bang predictions by Alpher, Herman and Gamow around 1950. Through the 1970s the radiation was found to be approximately consistent with a black body spectrum in all directions; this spectrum has been redshifted by the expansion of the universe, and today corresponds to approximately 2.725 K. This tipped the balance of evidence in favor of the Big Bang model, and Penzias and Wilson were awarded a Nobel Prize in 1978.

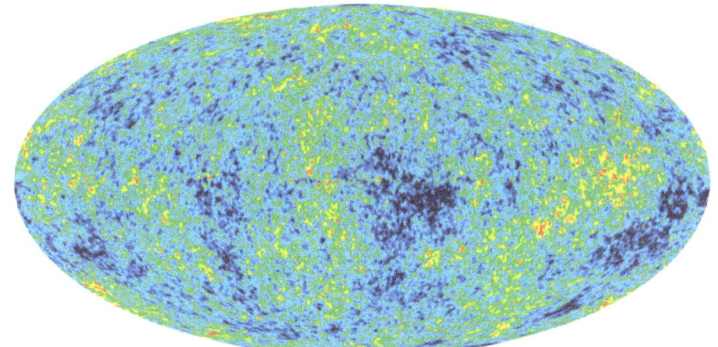

9 year WMAP image of the cosmic microwave background radiation (2012).
The radiation is isotropic to roughly one part in 100,000.

The *surface of last scattering* corresponding to emission of the CMB occurs shortly after *recombination*, the epoch when neutral hydrogen becomes stable. Prior to this, the universe comprised a hot dense photon-baryon plasma sea where photons were quickly scattered from free charged particles. Peaking at around 372 ± 14 kyr, the mean free path for a photon becomes long enough to reach the present day and the universe becomes transparent.

In 1989, NASA launched the Cosmic Background Explorer satellite (COBE), which made two major advances: in 1990, high-precision spectrum measurements showed, that the CMB frequency spectrum is an almost perfect blackbody with no deviations at a level of 1 part in 10^4, and measured a residual temperature of 2.726 K (more recent measurements have revised this figure down slightly to 2.7255 K); then in 1992, further COBE measurements discovered tiny fluctuations (anisotropies) in the CMB temperature across the sky, at a level of about one part in 10^5. John C. Mather and George Smoot were awarded the 2006 Nobel Prize in Physics for their leadership in these results.

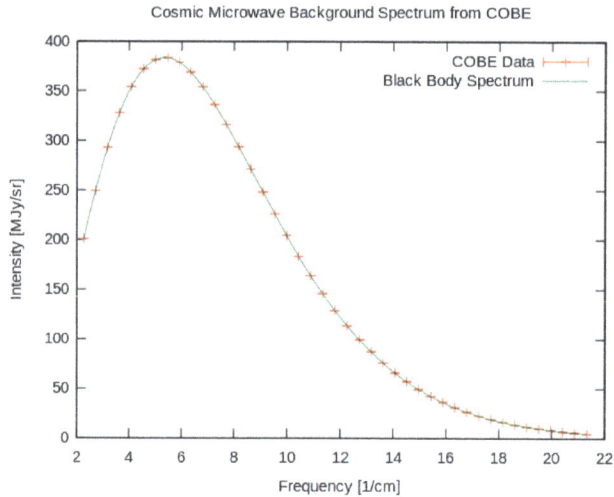

The cosmic microwave background spectrum measured by the FIRAS instrument on the COBE satellite is the most-precisely measured black body spectrum in nature. The data points and error bars on this graph are obscured by the theoretical curve.

During the following decade, CMB anisotropies were further investigated by a large number of ground-based and balloon experiments. In 2000–2001 several experiments, most notably BOO-MERanG, found the shape of the universe to be spatially almost flat by measuring the typical angular size (the size on the sky) of the anisotropies.

In early 2003, the first results of the Wilkinson Microwave Anisotropy Probe (WMAP) were released, yielding what were at the time the most accurate values for some of the cosmological parameters. The results disproved several specific cosmic inflation models, but are consistent with the inflation theory in general. The Planck space probe was launched in May 2009. Other ground and balloon based cosmic microwave background experiments are ongoing.

Abundance of Primordial Elements

Using the Big Bang model it is possible to calculate the concentration of helium-4, helium-3, deuterium, and lithium-7 in the universe as ratios to the amount of ordinary hydrogen. The relative abundances depend on a single parameter, the ratio of photons to baryons. This value can be calculated independently from the detailed structure of CMB fluctuations. The ratios predicted (by mass, not by number) are about 0.25 for $^4He/H$, about 10^{-3} for $^2H/H$, about 10^{-4} for $^3He/H$ and about 10^{-9} for $^7Li/H$.

The measured abundances all agree at least roughly with those predicted from a single value of the baryon-to-photon ratio. The agreement is excellent for deuterium, close but formally discrepant for ^4He, and off by a factor of two for ^7Li; in the latter two cases there are substantial systematic uncertainties. Nonetheless, the general consistency with abundances predicted by Big Bang nucleosynthesis is strong evidence for the Big Bang, as the theory is the only known explanation for the relative abundances of light elements, and it is virtually impossible to "tune" the Big Bang to produce much more or less than 20–30% helium. Indeed, there is no obvious reason outside of the Big Bang that, for example, the young universe (i.e., before star formation, as determined by studying matter supposedly free of stellar nucleosynthesis products) should have more helium than deuterium or more deuterium than ^{3}He, and in constant ratios, too.

Galactic Evolution and Distribution

Detailed observations of the morphology and distribution of galaxies and quasars are in agreement with the current state of the Big Bang theory. A combination of observations and theory suggest, that the first quasars and galaxies formed about a billion years after the Big Bang, and since then, larger structures have been forming, such as galaxy clusters and superclusters.

Populations of stars have been aging and evolving, so that distant galaxies (which are observed as they were in the early universe) appear very different from nearby galaxies (observed in a more recent state). Moreover, galaxies that formed relatively recently, appear markedly different from galaxies formed at similar distances but shortly after the Big Bang. These observations are strong arguments against the steady-state model. Observations of star formation, galaxy and quasar distributions and larger structures, agree well with Big Bang simulations of the formation of structure in the universe, and are helping to complete details of the theory.

Primordial Gas Clouds

In 2011, astronomers found what they believe to be pristine clouds of primordial gas by analyzing absorption lines in the spectra of distant quasars. Before this discovery, all other astronomical objects have been observed to contain heavy elements that are formed in stars. These two clouds of gas contain no elements heavier than hydrogen and deuterium. Since the clouds of gas have no heavy elements, they likely formed in the first few minutes after the Big Bang, during Big Bang nucleosynthesis.

Other Lines of Evidence

The age of the universe as estimated from the Hubble expansion and the CMB is now in good agreement with other estimates using the ages of the oldest stars, both as measured by applying the theory of stellar evolution to globular clusters and through radiometric dating of individual Population II stars.

The prediction that the CMB temperature was higher in the past has been experimentally supported by observations of very low temperature absorption lines in gas clouds at high redshift. This prediction also implies that the amplitude of the Sunyaev–Zel'dovich effect in clusters of galaxies does not depend directly on redshift. Observations have found this to be roughly true, but this effect depends on cluster properties that do change with cosmic time, making precise measurements difficult.

Future Observations

Future gravitational waves observatories might be able to detect primordial gravitational waves, relics of the early universe, up to less than a second after the Big Bang.

Problems and Related Issues in Physics

As with any theory, a number of mysteries and problems have arisen as a result of the development of the Big Bang theory. Some of these mysteries and problems have been resolved while others are still outstanding. Proposed solutions to some of the problems in the Big Bang model have revealed new mysteries of their own. For example, the horizon problem, the magnetic monopole problem, and the flatness problem are most commonly resolved with inflationary theory, but the details of the inflationary universe are still left unresolved and many, including some founders of the theory, say it has been disproven. What follows are a list of the mysterious aspects of the Big Bang theory still under intense investigation by cosmologists and astrophysicists.

Baryon Asymmetry

It is not yet understood why the universe has more matter than antimatter. It is generally assumed that when the universe was young and very hot, it was in statistical equilibrium and contained equal numbers of baryons and antibaryons. However, observations suggest that the universe, including its most distant parts, is made almost entirely of matter. A process called baryogenesis was hypothesized to account for the asymmetry. For baryogenesis to occur, the Sakharov conditions must be satisfied. These require that baryon number is not conserved, that C-symmetry and CP-symmetry are violated and that the universe depart from thermodynamic equilibrium. All these conditions occur in the Standard Model, but the effects are not strong enough to explain the present baryon asymmetry.

Dark Energy

Measurements of the redshift–magnitude relation for type Ia supernovae indicate, that the expansion of the universe has been accelerating since the universe was about half its present age. To explain this acceleration, general relativity requires, that much of the energy in the universe consists of a component with large negative pressure, dubbed "dark energy".

Dark energy, though speculative, solves numerous problems. Measurements of the cosmic microwave background indicate that the universe is very nearly spatially flat, and therefore according to general relativity the universe must have almost exactly the critical density of mass/energy. But the mass density of the universe can be measured from its gravitational clustering, and is found to have only about 30% of the critical density. Since theory suggests that dark energy does not cluster in the usual way it is the best explanation for the "missing" energy density. Dark energy also helps to explain two geometrical measures of the overall curvature of the universe, one using the frequency of gravitational lenses, and the other using the characteristic pattern of the large-scale structure as a cosmic ruler.

Negative pressure is believed to be a property of vacuum energy, but the exact nature and existence of dark energy remains one of the great mysteries of the Big Bang. Results from the WMAP team in 2008 are in accordance with a universe that consists of 73% dark energy, 23% dark matter, 4.6% regular matter and less than 1% neutrinos. According to theory, the energy density in matter decreases with the expansion of the universe, but the dark energy density remains constant (or nearly so) as the universe expands. Therefore, matter made up a larger fraction of the total energy of the universe in the past than it does today, but its fractional contribution will fall in the far future as dark energy becomes even more dominant.

The dark energy component of the universe has been explained by theorists using a variety of competing theories including Einstein's cosmological constant but also extending to more exotic forms of quintessence or other modified gravity schemes. A cosmological constant problem, sometimes called the "most embarrassing problem in physics", results from the apparent discrepancy between the measured energy density of dark energy, and the one naively predicted from Planck units.

Dark Matter

During the 1970s and the 1980s, various observations showed that there is not sufficient visible matter in the universe to account for the apparent strength of gravitational forces within and between galaxies. This led to the idea that up to 90% of the matter in the universe is dark matter that does not emit light or interact with normal baryonic matter. In addition, the assumption that the universe is mostly normal matter led to predictions that were strongly inconsistent with observations. In particular, the universe today is far more lumpy and contains far less deuterium than can be accounted for without dark matter. While dark matter has always been controversial, it is inferred by various observations: the anisotropies in the CMB, galaxy cluster velocity dispersions, large-scale structure distributions, gravitational lensing studies, and X-ray measurements of galaxy clusters.

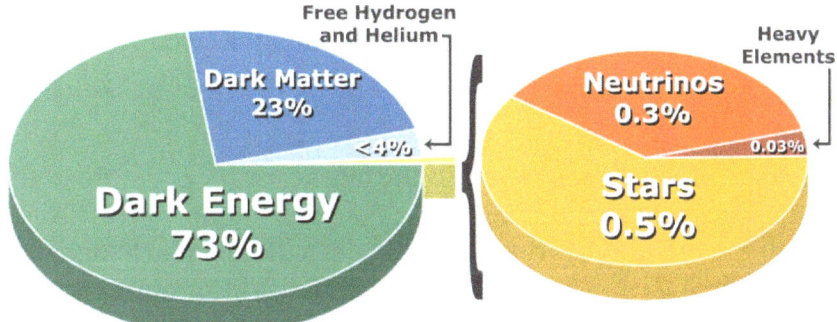

Chart shows the proportion of different components of the universe – about 95% is dark matter and dark energy.

Indirect evidence for dark matter comes from its gravitational influence on other matter, as no dark matter particles have been observed in laboratories. Many particle physics candidates for dark matter have been proposed, and several projects to detect them directly are underway.

Additionally, there are outstanding problems associated with the currently favored cold dark matter model which include the dwarf galaxy problem and the cuspy halo problem. Alternative theories have been proposed that do not require a large amount of undetected matter, but instead modify the laws of gravity established by Newton and Einstein; yet no alternative theory has been as successful as the cold dark matter proposal in explaining all extant observations.

Horizon Problem

The horizon problem results from the premise that information cannot travel faster than light. In a universe of finite age this sets a limit—the particle horizon—on the separation of any two regions of space that are in causal contact. The observed isotropy of the CMB is problematic in this regard: if the universe had been dominated by radiation or matter at all times up to the epoch of last scattering, the particle horizon at that time would correspond to about 2 degrees on the sky. There would then be no mechanism to cause wider regions to have the same temperature.

A resolution to this apparent inconsistency is offered by inflationary theory in which a homogeneous and isotropic scalar energy field dominates the universe at some very early period (before baryogenesis). During inflation, the universe undergoes exponential expansion, and the particle horizon expands much more rapidly than previously assumed, so that regions presently on opposite sides of the observable universe are well inside each other's particle horizon. The observed isotropy of the CMB then follows from the fact that this larger region was in causal contact before the beginning of inflation.

Heisenberg's uncertainty principle predicts that during the inflationary phase there would be quantum thermal fluctuations, which would be magnified to cosmic scale. These fluctuations serve as the seeds of all current structure in the universe. Inflation predicts that the primordial fluctuations are nearly scale invariant and Gaussian, which has been accurately confirmed by measurements of the CMB.

If inflation occurred, exponential expansion would push large regions of space well beyond our observable horizon.

A related issue to the classic horizon problem arises because in most standard cosmological inflation models, inflation ceases well before electroweak symmetry breaking occurs, so inflation should not be able to prevent large-scale discontinuities in the electroweak vacuum since distant parts of the observable universe were causally separate when the electroweak epoch ended.

Magnetic Monopoles

The magnetic monopole objection was raised in the late 1970s. Grand unified theories predicted topological defects in space that would manifest as magnetic monopoles. These objects would be produced efficiently in the hot early universe, resulting in a density much higher than is consistent with observations, given that no monopoles have been found. This problem is also resolved by cosmic inflation, which removes all point defects from the observable universe, in the same way that it drives the geometry to flatness.

Flatness Problem

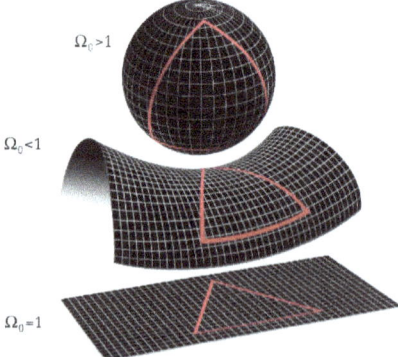

The overall geometry of the universe is determined by whether the Omega cosmological parameter is less than, equal to or greater than 1. Shown from top to bottom are a closed universe with positive curvature, a hyperbolic universe with negative curvature and a flat universe with zero curvature.

The flatness problem (also known as the oldness problem) is an observational problem associated with a Friedmann–Lemaître–Robertson–Walker metric (FLRW). The universe may have positive, negative, or zero spatial curvature depending on its total energy density. Curvature is negative, if its density is less than the critical density; positive, if greater; and zero at the critical density, in which case space is said to be *flat*.

The problem is, that any small departure from the critical density grows with time, and yet the universe today remains very close to flat. Given that a natural timescale for departure from flatness might be the Planck time, 10^{-43} seconds, the fact that the universe has reached neither a heat death nor a Big Crunch after billions of years requires an explanation. For instance, even at the relatively late age of a few minutes (the time of nucleosynthesis), the universe density must have been within one part in 10^{14} of its critical value, or it would not exist as it does today.

Origin of Conditions Required for Big Bang to Occur

Gottfried Wilhelm Leibniz wrote: *"Why is there something rather than nothing? The sufficient reason [...] is found in a substance which [...] is a necessary being bearing the reason for its existence within itself."* Philosopher of physics Dean Rickles has argued that numbers and mathematics (or their underlying laws) may necessarily exist. Quantum fluctuation, or other physics laws that may have existed at the start of the Big Bang, could then create the conditions for matter to occur.

Ultimate Fate of the Universe

Before observations of dark energy, cosmologists considered two scenarios for the future of the universe. If the mass density of the universe were greater than the critical density, then the universe would reach a maximum size and then begin to collapse. It would become denser and hotter again, ending with a state similar to that in which it started—a Big Crunch.

Alternatively, if the density in the universe were equal to—or below the critical density, the expansion would slow down but never stop. Star formation would cease with the consumption of interstellar gas in each galaxy; stars would burn out leaving white dwarfs, neutron stars, and black holes. Very gradually, collisions between these would result in mass accumulating into larger and larger black holes. The average temperature of the universe would asymptotically approach absolute zero—a Big Freeze. Moreover, if the proton were unstable, then baryonic matter would disappear, leaving only radiation and black holes. Eventually, black holes would evaporate by emitting Hawking radiation. The entropy of the universe would increase to the point, where no organized form of energy could be extracted from it; a scenario known as heat death.

Modern observations of accelerating expansion imply that more and more of the currently visible universe will pass beyond our event horizon and out of contact with us. The eventual result is not known. The ΛCDM model of the universe contains dark energy in the form of a cosmological constant. This theory suggests that only gravitationally bound systems, such as galaxies, will remain together, and they too will be subject to heat death as the universe expands and cools. Other explanations of dark energy, called phantom energy theories, suggest that ultimately galaxy clusters, stars, planets, atoms, nuclei, and matter itself will be torn apart by the ever-increasing expansion in a so-called Big Rip.

Speculations

While the Big Bang model is well established in cosmology, it is likely to be refined. The Big Bang theory, built upon the equations of classical general relativity, indicates a singularity at the origin of cosmic time; this infinite energy density is regarded as impossible in physics. Still, it is known that the equations are not applicable before the time when the universe cooled down to the Planck temperature, and this conclusion depends on various assumptions, of which some could never be experimentally verified.

One proposed refinement to avoid this would-be singularity is to develop a correct treatment of quantum gravity.

It is not known what could have preceded the hot dense state of the early universe or how and why it originated, though speculation abounds in the field of cosmogony.

Some proposals, each of which entails untested hypotheses, are:

- Models including the Hartle–Hawking no-boundary condition, in which the whole of space-time is finite; the Big Bang does represent the limit of time but without any singularity.

- Big Bang lattice model, states that the universe at the moment of the Big Bang consists of an infinite lattice of fermions, which is smeared over the fundamental domain so it has rotational, translational and gauge symmetry. The symmetry is the largest symmetry possible and hence the lowest entropy of any state.

- Brane cosmology models, in which inflation is due to the movement of branes in string theory; the pre-Big Bang model; the ekpyrotic model, in which the Big Bang is the result of a collision between branes; and the cyclic model, a variant of the ekpyrotic model in which collisions occur periodically. In the latter model the Big Bang was preceded by a Big Crunch and the universe cycles from one process to the other.

- Eternal inflation, in which universal inflation ends locally here and there in a random fashion, each end-point leading to a *bubble universe*, expanding from its own big bang.

Religious and Philosophical Interpretations

As a description of the origin of the universe, the Big Bang has significant bearing on religion and philosophy. As a result, it has become one of the liveliest areas in the discourse between science and religion. Some believe the Big Bang implies a creator, and some see its mention in their holy books, while others argue that Big Bang cosmology makes the notion of a creator superfluous.

Lambda-CDM Model

The ΛCDM (Lambda cold dark matter) or Lambda-CDM model is a parametrization of the Big Bang cosmological model in which the universe contains a cosmological constant, denoted by Lambda (Greek Λ), associated with dark energy, and cold dark matter (abbreviated CDM). It is frequently referred to as the standard model of Big Bang cosmology because it is the simplest model that provides a reasonably good account of the following properties of the cosmos:

- the existence and structure of the cosmic microwave background

- the large-scale structure in the distribution of galaxies

- the abundances of hydrogen (including deuterium), helium, and lithium

- the accelerating expansion of the universe observed in the light from distant galaxies and supernovae

The model assumes that general relativity is the correct theory of gravity on cosmological scales. It emerged in the late 1990s as a concordance cosmology, after a period of time when disparate observed properties of the universe appeared mutually inconsistent, and there was no consensus on the makeup of the energy density of the universe.

The ΛCDM model can be extended by adding cosmological inflation, quintessence and other elements that are current areas of speculation and research in cosmology.

Some alternative models challenge the assumptions of the ΛCDM model. Examples of these are modified Newtonian dynamics, modified gravity and theories of large-scale variations in the matter density of the universe.

Overview

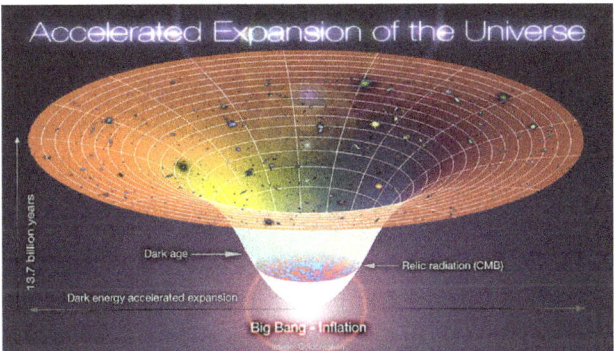

Lambda-CDM, accelerated expansion of the universe. The time-line in this schematic diagram extends from the Big Bang/inflation era 13.7 Gyr ago to the present cosmological time.

Most modern cosmological models are based on the cosmological principle, which states that our observational location in the universe is not unusual or special; on a large-enough scale, the universe looks the same in all directions (isotropy) and from every location (homogeneity).

The model includes an expansion of metric space that is well documented both as the red shift of prominent spectral absorption or emission lines in the light from distant galaxies and as the time dilation in the light decay of supernova luminosity curves. Both effects are attributed to a Doppler shift in electromagnetic radiation as it travels across expanding space. Although this expansion increases the distance between objects that are not under shared gravitational influence, it does not increase the size of the objects (e.g. galaxies) in space. It also allows for distant galaxies to recede from each other at speeds greater than the speed of light; local expansion is less than the speed of light, but expansion summed across great distances can collectively exceed the speed of light.

The letter Λ (lambda) represents the cosmological constant, which is currently associated with a vacuum energy or dark energy in empty space that is used to explain the contemporary accelerating expansion of space against the attractive effects of gravity. A cosmological constant has negative pressure, $p = -\rho c^2$, which contributes to the stress-energy tensor that, according to the general theory of relativity, causes accelerating expansion. The fraction of the total energy density of our (flat or almost flat) universe that is dark energy, Ω_Λ, is currently estimated to be 0.692 ± 0.012, or even 0.6911 ± 0.0062 based on Planck satellite data.

Cold dark matter is a form of matter introduced in order to account for gravitational effects observed in very large-scale structures (the "flat" rotation curves of galaxies; the gravitational lensing of light by galaxy clusters; and enhanced clustering of galaxies) that cannot be accounted for by the quantity of observed matter. Dark matter is described as being cold (i.e. its velocity is far less than the speed of light at the epoch of radiation-matter equality); non-baryonic (i.e. consisting of matter other than protons and neutrons); dissipationless (i.e. cannot cool by radiating photons); and collisionless (i.e. the dark matter particles interact with each other and other particles only through gravity and possibly the weak force). The dark matter component is currently estimated to constitute about 26.8% of the mass-energy density of the universe.

The remaining 4.9% comprises all ordinary matter observed as atoms, chemical elements, gas and plasma, the stuff of which visible planets, stars and galaxies are made. As a matter of fact, the great majority of ordinary matter in the universe is unseen, since visible stars and gas inside galaxies and clusters account for less than 10 per cent of the ordinary matter contribution to the mass-energy density of the universe.

Also, the energy density includes a very small fraction (~ 0.01%) in cosmic microwave background radiation, and not more than 0.5% in relic neutrinos. Although very small today, these were much more important in the distant past, dominating the matter at redshift > 3200.

The model includes a single originating event, the "Big Bang", which was not an explosion but the abrupt appearance of expanding space-time containing radiation at temperatures of around 10^{15} K. This was immediately (within 10^{-29} seconds) followed by an exponential expansion of space by a scale multiplier of 10^{27} or more, known as cosmic inflation. The early universe remained hot (above 10,000 K) for several hundred thousand years, a state that is detectable as a residual cosmic microwave background, or CMB, a very low energy radiation emanating from all parts of the sky. The "Big Bang" scenario, with cosmic inflation and standard particle physics, is the only current cosmological model consistent with the observed continuing expansion of space, the observed distribution of lighter elements in the universe (hydrogen, helium, and lithium), and the spatial texture of minute irregularities (anisotropies) in the CMB radiation. Cosmic inflation also addresses the "horizon problem" in the CMB; indeed, it seems likely that the universe is larger than the observable particle horizon.

The model uses the FLRW metric, the Friedmann equations and the cosmological equations of state to describe the observable universe from right after the inflationary epoch to present and future.

Cosmic Expansion History

The expansion of the universe is parametrized by a dimensionless scale factor $a = a(t)$ (with time t counted from the birth of the universe), defined relative to the present day, so $a_0 = a(t_0) = 1$; the usual convention in cosmology is that subscript 0 denotes present-day values, so t_0 is the current age of the universe. The scale factor is related to the observed redshift z of the light emitted at time t_{em} by

$$\frac{1}{a(t_{em})} = 1 + z.$$

The expansion rate is described by the time-dependent Hubble parameter, $H(t)$, defined as

$$H(t) \equiv \frac{\dot{a}}{a},$$

where $\dot{a}$ is the time-derivative of the scale factor. The first Friedmann equation gives the expansion rate in terms of the matter+radiation density ρ, the curvature k, the cosmological constant Λ, and the gravitational constant :

$$H^2 = \left(\frac{\dot{a}}{a}\right)^2 = \frac{8\pi G}{3}\rho - \frac{kc^2}{a^2} + \frac{\Lambda c^2}{3}.$$

A critical density ρ_{crit} is the present-day density, which gives zero curvature k, assuming the cosmological constant Λ is zero, regardless of its actual value. Substituting these conditions to the Friedmann equation gives

$$\rho_{\text{crit}} = \frac{3H_0^2}{8\pi G} = 1.878\ 47(23)\times 10^{-26}\ h^2\ \text{kg m}^{-3}$$

where $h \equiv H_0 / (100\ \text{km s}^{-1}\ \text{Mpc}^{-1})$ is the reduced Hubble constant. If the cosmological constant were actually zero, the critical density would also be the dividing line between eventual recollapse of the universe to a Big Crunch, or unlimited expansion. Because it is not, the universe is predicted to expand forever regardless of whether the total density is slightly above or below the critical density, though this may not apply if the cosmological constant is time-dependent.

It is standard to define the present-day density parameter Ω_x for various species as the dimensionless ratio

$$\Omega_x \equiv \frac{\rho_x(t=t_0)}{\rho_{\text{crit}}} = \frac{8\pi G \rho_x(t=t_0)}{3H_0^2}$$

where the subscript x is one of "b" for baryons, "c" for cold dark matter, "rad" for radiation (photons plus relativistic neutrinos), and "DE" or "Λ" for dark energy.

Since the densities of various species scale as different powers of a, e.g. a^{-3} for matter etc., the Friedmann equation can be conveniently rewritten in terms of the various density parameters as

$$H(a) \equiv \frac{\dot{a}}{a} = H_0\sqrt{(\Omega_c + \Omega_b)a^{-3} + \Omega_{rad}a^{-4} + \Omega_k a^{-2} + \Omega_{DE}a^{-3(1+w)}}$$

where w is the equation of state of dark energy, and assuming negligible neutrino mass (significant neutrino mass requires a more complex equation). The various Ω parameters add up to 1 by construction. In the general case this is integrated by computer to give the expansion history a(t) and also observable distance-redshift relations for any chosen values of the cosmological parameters, which can then be compared with observations such as supernovae and baryon acoustic oscillations.

In the minimal 6-parameter Lambda-CDM model, it is assumed that curvature Ω_k is zero and $w = -1$, so this simplifies to

$$H(a) = H_0 \sqrt{\Omega_m a^{-3} + \Omega_{rad} a^{-4} + \Omega_\Lambda}$$

Observations show that the radiation density is very small today, $\Omega_{rad} \sim 10^{-4}$; if this term is neglected the above has an analytic solution

$$a(t) = (\Omega_m / \Omega_\Lambda)^{1/3} \sinh^{2/3}(t / t_\Lambda)$$

where $t_\Lambda \equiv 2 / (3H_0 \sqrt{\Omega_\Lambda})$ this is fairly accurate for a > 0.01 or t > 10 Myr. Solving for $a(t) = 1$ gives the present age of the universe t_0 in terms of the other parameters.

It follows that the transition from decelerating to accelerating expansion (the second derivative $\ddot{a}$ crossing zero) occurred when

$$a = (\Omega_m / 2\Omega_\Lambda)^{1/3}$$

which evaluates to a ~ 0.6 or z ~ 0.66 for the Planck best-fit parameters.

Historical Development

The discovery of the Cosmic Microwave Background (CMB) in 1965 confirmed a key prediction of the Big Bang cosmology. From that point on, it was generally accepted that the universe started in a hot, dense state and has been expanding over time. The rate of expansion depends on the types of matter and energy present in the universe, and in particular, whether the total density is above or below the so-called critical density. During the 1970s, most attention focused on pure-baryonic models, but there were serious challenges explaining the formation of galaxies, given the small anisotropies in the CMB (upper limits at that time). In the early 1980s, it was realized that this could be resolved if cold dark matter dominated over the baryons, and the theory of cosmic inflation motivated models with critical density. During the 1980s, most research focused on cold dark matter with critical density in matter, around 95% CDM and 5% baryons: these showed success at forming galaxies and clusters of galaxies, but problems remained; notably, the model required a Hubble constant lower than preferred by observations, and observations around 1988-1990 showed more large-scale galaxy clustering than predicted. These difficulties sharpened with the discovery of CMB anisotropy by COBE in 1992, and several modified CDM models, including ΛCDM and mixed cold+hot dark matter, came under active consideration through the mid-1990s. The ΛCDM model then became the leading model following the observations of accelerating expansion in 1998, and was quickly supported by other observations: in 2000, the BOOMERanG microwave background experiment measured the total (matter+energy) density to be close to 100% of critical, whereas in 2001 the 2dFGRS galaxy redshift survey measured the matter density to be near 25%; the large difference between these values supports a positive Λ or dark energy. Much more precise spacecraft measurements of the microwave background from WMAP in 2003 – 2010 and Planck in 2013 - 2015 have continued to support the model and pin down the parameter values, most of which are now constrained below 1 percent uncertainty.

There is currently active research into many aspects of the ΛCDM model, both to refine the parameters and possibly detect deviations. In addition, ΛCDM has no explicit physical theory for the origin or physical nature of dark matter or dark energy; the nearly scale-invariant spectrum of the CMB perturbations, and their image across the celestial sphere, are believed to result from very small thermal and acoustic irregularities at the point of recombination. A large majority of astronomers and astrophysicists support the ΛCDM model or close relatives of it, but Milgrom, McGaugh, and Kroupa are leading critics, attacking the dark matter portions of the theory from the perspective of galaxy formation models and supporting the alternative MOND theory, which requires a modification of the Einstein field equations and the Friedmann equations as seen in proposals such as MOG theory or TeVeS theory. Other proposals by theoretical astrophysicists of cosmological alternatives to Einstein's general relativity that attempt to account for dark energy or dark matter include f(R) gravity, scalar–tensor theories such as galileon theories, brane cosmologies, the DGP model, and massive gravity and its extensions such as bimetric gravity.

Successes

In addition to explaining pre-2000 observations, the model has made a number of successful predictions: notably the existence of the baryon acoustic oscillation feature, discovered in 2005 in the predicted location; and the statistics of weak gravitational lensing, first observed in 2000 by several teams. The polarization of the CMB, discovered in 2002 by DASI is now a dramatic success: in the 2015 Planck data release, there are seven observed peaks in the temperature (TT) power spectrum, six peaks in the temperature-polarization (TE) cross spectrum, and five peaks in the polarization (EE) spectrum. The six free parameters can be well constrained by the TT spectrum alone, and then the TE and EE spectra can be predicted theoretically to few-percent precision with no further adjustments allowed: comparison of theory and observations shows an excellent match.

Challenges

Extensive searches for dark matter particles have so far shown no well-agreed detection; the dark energy may be almost impossible to detect in a laboratory, and its value is unnaturally small compared to naive theoretical predictions.

Comparison of the model with observations is very successful on large scales (larger than galaxies, up to the observable horizon), but may have some problems on sub-galaxy scales, possibly predicting too many dwarf galaxies and too much dark matter in the innermost regions of galaxies. These small scales are harder to resolve in computer simulations, so it is not yet clear whether the problem is the simulations, non-standard properties of dark matter, or a more radical error in the model.

It has been argued that the ΛCDM model is built upon a foundation of conventionalist stratagems, rendering it unfalsifiable in the sense defined by Karl Popper.

Parameters

The simple ΛCDM model is based on six parameters: physical baryon density parameter; physical dark matter density parameter; the age of the universe; scalar spectral index; curvature fluctuation amplitude; and reionization optical depth. In accordance with Occam's razor, six is the smallest number of parameters needed to give an acceptable fit to current observations; other possible parameters are fixed at "natural" values, e.g. total density parameter = 1.00, dark energy equation of state = −1.

The values of these six parameters are mostly not predicted by current theory (though, ideally, they may be related by a future "Theory of Everything"), except that most versions of cosmic inflation predict the scalar spectral index should be slightly smaller than 1, consistent with the estimated value 0.96. The parameter values, and uncertainties, are estimated using large computer searches to locate the region of parameter space providing an acceptable match to cosmological observations. From these six parameters, the other model values, such as the Hubble constant and the dark energy density, can be readily calculated.

Commonly, the set of observations fitted includes the cosmic microwave background anisotropy, the brightness/redshift relation for supernovae, and large-scale galaxy clustering including the baryon acoustic oscillation feature. Other observations, such as the Hubble constant, the abundance of galaxy clusters, weak gravitational lensing and globular cluster ages, are generally consistent with these, providing a check of the model, but are less precisely measured at present.

Parameter values listed below are from the Planck Collaboration Cosmological parameters 68% confidence limits for the base ΛCDM model from Planck CMB power spectra, in combination with lensing reconstruction and external data (BAO+JLA+H_0).

1. The "physical baryon density parameter" $\Omega_b h^2$ is the "baryon density parameter" Ω_b multiplied by the square of the reduced Hubble constant $h = H_0 / (100 \text{ km s}^{-1} \text{ Mpc}^{-1})$. Likewise for the difference between "physical dark matter density parameter" and "dark matter density parameter".

2. A density $\rho_{x} = \Omega_x \rho_{crit}$ is expressed in terms of the critical density ρcrit, which is the total density of matter/energy needed for the universe to be spatially flat. Measurements indicate that the actual total density ρ_{tot} is very close if not equal to this value.

3. This is the minimal value allowed by solar and terrestrial neutrino oscillation experiments.

4. from the Standard Model of particle physics

5. Calculated from $\Omega_{bh}{}^2$ and $h = H_0 / (100 \text{ km s}^{-1} \text{ Mpc}^{-1})$.

6. Calculated from $\Omega_{ch}{}^2$ and $h = H_0 / (100 \text{ km s}^{-1} \text{ Mpc}^{-1})$.

7. Calculated from $h = H_0 / (100 \text{ km s}^{-1} \text{ Mpc}^{-1})$ per $\rho_{crit} = 1.87847 \times 10^{-26} \, h^2 \text{ kg m}^{-3}$.

Planck Collaboration Cosmological parameters			
	Description	Symbol	Value
	Physical baryon density parameter	$\Omega_b h^2$	0.02230 ± 0.00014
	Physical dark matter density parameter	$\Omega_c h^2$	0.1188 ± 0.0010
Independent parameters	Age of the universe	t_0	$13.799 \pm 0.021 \times 10^9$ years
	Scalar spectral index	n_s	0.9667 ± 0.0040
	Curvature fluctuation amplitude, $k_0 = 0.002 \text{ Mpc}^{-1}$	$\Delta 2$ R	$2.441 + 0.088$ -0.092×10^{-9}
	Reionization optical depth	τ	0.066 ± 0.012

Fixed parameters	Total density parameter	Ω_{tot}	1
	Equation of state of dark energy	w	−1
	Sum of three neutrino masses	Σm_ν	0.06 eV/c²
	Effective number of relativistic degrees of freedom	N_{eff}	3.046
	Tensor/scalar ratio	r	0
	Running of spectral index	$d\,n_s / d\ln k$	0
Calculated values	Hubble constant	H_0	67.74±0.46 km s⁻¹ Mpc⁻¹
	Baryon density parameter	Ω_b	0.0486±0.0010
	Dark matter density parameter	Ω_c	0.2589±0.0057
	Matter density parameter	Ω_m	0.3089±0.0062
	Dark energy density parameter	Ω_Λ	0.6911±0.0062
	Critical density	ρ_{crit}	(8.62±0.12)×10⁻²⁷ kg/m³
	Fluctuation amplitude at 8h⁻¹ Mpc	σ_8	0.8159±0.0086
	Redshift at decoupling	z_*	1089.90±0.23
	Age at decoupling	t_*	377700±3200 years
	Redshift of reionization (with uniform prior)	z_{re}	8.5+1.0−1.1

Extended Models

Extended model parameters		
Description	**Symbol**	**Value**
Total density parameter	Ω_{tot}	1.0023+0.0056−0.0054
Equation of state of dark energy	w	−0.980±0.053
Tensor-to-scalar ratio	r	< 0.11, k_0 = 0.002 Mpc⁻¹ (2σ)
Running of the spectral index	$d\,n_s / d\ln k$	−0.022±0.020, k_0 = 0.002 Mpc⁻¹
Physical neutrino density parameter	$\Omega_\nu h^2$	< 0.0062
Sum of three neutrino masses	$\Sigma\,m_\nu$	< 0.58 eV/c² (2σ)

Extended models allow one or more of the "fixed" parameters above to vary, in addition to the basic six; so these models join smoothly to the basic six-parameter model in the limit that the additional parameter(s) approach the default values. For example, possible extensions of the simplest ΛCDM model allow for spatial curvature (Ω_{tot} may be different from 1); or quintessence rather than a cosmological constant where the equation of state of dark energy is allowed to differ from −1. Cosmic inflation predicts tensor fluctuations (gravitational waves). Their amplitude is parameterized by the tensor-to-scalar ratio (denoted *r*), which is determined by the unknown energy scale of inflation. Other modifications allow hot dark matter in the form of neutrinos more massive than the minimal value, or a running spectral index; the latter is generally not favoured by simple cosmic inflation models.

Allowing additional variable parameter(s) will generally *increase* the uncertainties in the standard six parameters quoted above, and may also shift the central values slightly. The Table shows results for each of the possible "6+1" scenarios with one additional variable parameter; this indicates that, as of 2015, there is no convincing evidence that any additional parameter is different from its default value.

Some researchers have suggested that there is a running spectral index, but no statistically significant study has revealed one. Theoretical expectations suggest that the tensor-to-scalar ratio r should be between 0 and 0.3, and the latest results are now within those limits.

World Egg

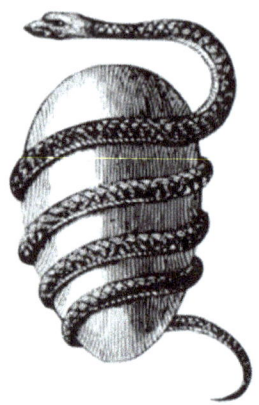

Jacob Bryant's Orphic Egg (1774)

The world egg, cosmic egg or mundane egg is a mythological motif found in the creation myths of many cultures and civilizations. Typically, the world egg is a beginning of some sort, and the universe or some primordial being comes into existence by "hatching" from the egg, sometimes lain on the primordial waters of the Earth.

Vedic Mythology

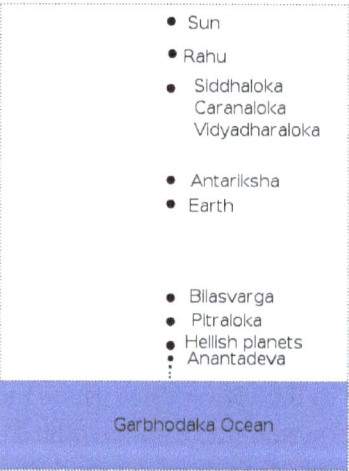

Vivasvan, Rahu, Bhūmi, Naraka, Ananta, Garbhodaksayi Vishnu

One of the earliest ideas of "egg-shaped cosmos" comes from some of the Sanskrit scriptures. The Sanskrit term for it is Brahmanda which is derived from two words- 'Brahm' means 'cosmos' or 'expanding' and 'anda' means 'egg'. Certain Puranas such as the Brahmanda Purana speak of this in detail.

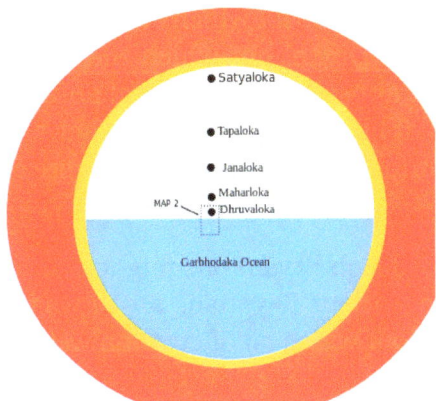

This is one of many material universes, Brahmandas, which expand from Mahavishnu when He breathes.

The Rig Veda (RV 10.121) uses a similar name for the source of the universe: Hiranyagarbha which literally means "golden fetus" or "golden womb". The Upanishads elaborate that the Hiranyagarbha floated around in emptiness for a while, and then broke into two halves which formed Dyaus (Heaven) and Prithvi (Earth). The Rig Veda has a similar coded description of the division of the universe in its early stages.

Greek Mythology

The Orphic Egg in the ancient Greek Orphic tradition is the cosmic egg from which hatched the primordial hermaphroditic deity Phanes/Protogonus (variously equated also with Zeus, Pan, Metis, Eros, Erikepaios and Bromius) who in turn created the other gods. The egg is often depicted with a serpent wound around it.

Many threads of earlier myths are apparent in the new tradition. Phanes was believed to have been hatched from the World-Egg of Chronos (Time) and Ananke (Necessity). His older wife Nyx (Night) called him Protogenus. As she created nighttime, he created daytime. He also created the method of creation by mingling. He was made the ruler of the deities and passed the sceptre to Nyx. This new Orphic tradition states that Nyx later gave the sceptre to her son Uranos before it passed to Cronus and then to Zeus, who retained it.

Egyptian Mythology

The Ancient Egyptians abided by a multiplicity of truths when it came to creation myths. For instance, the Hermopolitan, Heliopolitan and Memphite theologies, were equally validated. Under the Hermopolitan theology, there is the Ogdoad, which represents the conditions before the Gods were created Van Dijk, 1995). An aspect within the Ogdoad is the Cosmic Egg, from which all things are born. Life comes from the Cosmic Egg; the sun god Re was born from the primordial egg in a stage known as the first occasion (Dunand, 2004).

Phoenician Mythology

A philosophical creation story traced to "the cosmogony of *Taautus*, whom Philo of Byblos explicitly identified with the Egyptian Thoth—"the first who thought of the invention of letters, and began the writing of records"— which begins with Erebus and Wind, between which *Eros* 'Desire' came to be. From this was produced *Môt* which seems to be the Phoenician/Ge'ez/Hebrew/Arabic/Ancient Egyptian word for 'Death' but which the account says may mean 'mud'. In a mixed confusion, the germs of life appear, and intelligent animals called *Zophasemin* (explained probably correctly as 'observers of heaven') formed together as an egg, perhaps. The account is not clear. Then Môt burst forth into light and the heavens were created and the various elements found their stations.

Following the etymological line of Jacob Bryant one might also consider with regard to the meaning of *Môt*, that according to the Ancient Egyptians *Ma'at* was the personification of the fundamental order of the universe, without which all of creation would perish. She was also considered the wife of Thoth.

Chinese Mythology

In the myth of Pangu, developed by Taoist monks hundreds of years after Lao Zi, the universe began as an egg. A god named Pangu, born inside the egg, broke it into two halves: the upper half became the sky, while the lower half became the earth. As the god grew taller, the sky and the earth grew thicker and were separated further. Finally Pangu died and his body parts became different parts of the earth.

Finnish Mythology

In the *Kalevala*, the Finnish national epic, there is a myth of the world being created from the fragments of an egg laid by a diving duck on the knee of Ilmatar, goddess of the air:

> *One egg's lower half transformed*
>
> *And became the earth below,*
>
> *And its upper half transmuted*
>
> *And became the sky above;*
>
> *From the yolk the sun was made,*
>
> *Light of day to shine upon us;*
>
> *From the white the moon was formed,*
>
> *Light of night to gleam above us;*
>
> *All the colored brighter bits*
>
> *Rose to be the stars of heaven*
>
> *And the darker crumbs changed into*
>
> *Clouds and cloudlets in the sky.*

In many original folk poems, the duck - or sometimes an eagle - laids its eggs on the knee of Väinämöinen.

Polynesian Mythology

In Cook Islands mythology, deep within Avaiki (the Underworld), a place described as resembling a vast hollow coconut shell, there dwelt in the deepest depths, the primordial mother goddess, Varima-te-takere. Her domain was described as being so narrow, that her knees touched her chin. It was from this place that she created the first man, Avatea, a god of light, a hybrid being half man and half fish. He was sent to the Upperworld to shine light in the land of men, and his eyes were believed to be the sun and the moon.

Representations

- In the temple of Daibod, Japan, it is represented as a nest egg floating in an expanse of water.

- On the island of Cyprus, the egg is represented as a gigantic egg-shaped vase.

Modern Mythology

In 1955 poet and writer Robert Graves published the mythography *The Greek Myths*, a compendium of Greek mythology normally published in two volumes. Within this work Graves' imaginatively reconstructed "Pelasgian creation myth" features a supreme creatrix, Eurynome, "The Goddess of All Things", who arose naked from Chaos to part sea from sky so that she could dance upon the waves. Catching the north wind at her back and, rubbing it between her hands, she warms the *pneuma* and spontaneously generates the serpent Ophion, who mates with her. In the form of a dove upon the waves, she lays the Cosmic Egg and bids Ophion to incubate it by coiling seven times around until it splits in two and hatches "all things that exist... sun, moon, planets, stars, the earth with its mountains and rivers, its trees, herbs, and living creatures".

In Modern Cosmology

The concept was resurrected by modern science in the 1930s and explored by theoreticians during the following two decades. The idea comes from a perceived need to reconcile Edwin Hubble's observation of an expanding universe (which was also predicted from Einstein's equations of general relativity by Alexander Friedmann) with the notion that the universe must be eternally old. Current cosmological models maintain that 13.8 billion years ago, the entire mass of the universe was compressed into a gravitational singularity, the so-called cosmic egg, from which it expanded to its current state (following the Big Bang).

Georges Lemaitre proposed in 1927 that the cosmos originated from what he called the *primeval atom.*

In the late 1940s, George Gamow's assistant cosmological researcher Ralph Alpher, proposed the name ylem for the primordial substance that existed between the big crunch of the previous universe and the big bang of our own universe.

References

- Öpik, E. (1922). "An estimate of the distance of the Andromeda Nebula". Astrophysical Journal. 55: 406–410. Bibcode:1922ApJ....55..406O. doi:10.1086/142680

- Linton, C.M. (2004). From Eudoxus to Einstein: A History of Mathematical Astronomy. E-Libro. Cambridge University Press. p. 38. ISBN 9781139453790

- Wollack, Edward J. (10 December 2010). "Cosmology: The Study of the Universe". Universe 101: Big Bang Theory. NASA. Retrieved 27 April 2011

- Glanz, James (1998). "Breakthrough of the year 1998. Astronomy: Cosmic Motion Revealed". Science. 282 (5397): 2156–2157. Bibcode:1998Sci...282.2156G. doi:10.1126/science.282.5397.2156a

- Eastwood, B. S. (1992-11-01). "Heraclides and heliocentrism – Texts diagrams and interpretations". Journal for the History of Astronomy. 23: 233. Bibcode:1992JHA....23..233E

- Hwang, Helen Hye-Sook (2015). The Mago Way: Re-discovering Mago, the Great Goddess from East Asia. Mago Books. pp. 136–142. ISBN 9781516907922

- Livio, Mario (2001). The Accelerating Universe: Infinite Expansion, the Cosmological Constant, and the Beauty of the Cosmos. John Wiley and Sons. p. 53. Retrieved 31 March 2012

- "Parallel Universes". Scientific American. 288: 40–51. May 2003. Bibcode:2003SciAm.288e..40T. PMID 12701329. arXiv:astro-ph/0302131. doi:10.1038/scientificamerican0503-40

- Loeb, Abraham (October 2014). "The Habitable Epoch of the Early Universe". International Journal of Astrobiology. 13 (4): 337–339. Bibcode:2014IJAsB..13..337L. arXiv:1312.0613. doi:10.1017/S1473550414000196

- Lawson, Russell M. (2004). Science in the Ancient World: An Encyclopedia. ABC-CLIO. pp. 29–30. ISBN 1851095349

- Babinski, E. T., ed. (1995). "Excerpts from Frank Zindler's 'Report from the center of the universe' and 'Turtles all the way down'". TalkOrigins Archive. Retrieved 2013-12-01

- Curtis, H. D. (1988). "Novae in Spiral Nebulae and the Island Universe Theory". Publications of the Astronomical Society of the Pacific. 100: 6. Bibcode:1988PASP..100....6C. doi:10.1086/132128

- Spergel, D. N.; et al. (2003). "First year Wilkinson Microwave Anisotropy Probe (WMAP) observations: determination of cosmological parameters". The Astrophysical Journal Supplement. 148 (1): 175–194

- Kirmani, M. Zaki; Singh, Nagendra Kr (2005). Encyclopaedia of Islamic Science and Scientists: A-H. Global Vision. ISBN 9788182200586

- DeYoung, Donald B. (1997-11-05). "Astronomy and the Bible: Selected questions and answers excerpted from the book". Answers in Genesis. Retrieved 2013-12-01

- Ragep, F. Jamil (2001). "Tusi and Copernicus: The Earth's motion in context". Science in Context. Cambridge University Press. 14 (1-2): 145–163. doi:10.1017/s0269889701000060

- Penzias, A. A.; Wilson, R. W. (1965). "A Measurement of Excess Antenna Temperature at 4080 Mc/s". The Astrophysical Journal. 142: 419. Bibcode:1965ApJ...142..419P. doi:10.1086/148307

- Hoskin, Michael (1999-03-18). The Cambridge Concise History of Astronomy. Cambridge University Press. p. 60. ISBN 9780521576000

- Nussbaum, Alexander (2007-12-19). "Orthodox Jews & science: An empirical study of their attitudes toward evolution, the fossil record, and modern geology". Skeptic Magazine. Retrieved 2008-12-18

- Ragep, F. Jamil (2001). "Freeing astronomy from philosophy: An aspect of Islamic influence on science". Osiris. 2nd Series. 16 (Science in Theistic Contexts: Cognitive Dimensions): 49–64, 66–71. doi:10.1086/649338

- Melchiorri, A., et. al. (1999). "A measurement of Omega from the North American test flight of BOOMERANG". The Astrophysical Journal. Institute of Physics. 536. arXiv:astro-ph/9911445. doi:10.1086/312744

- Huff, Toby E. (2003). The Rise of Early Modern Science: Islam, China and the West. Cambridge University Press. p. 58. ISBN 9780521529945

An Overview of Physical Cosmology

Physical cosmology is the study of the structure of the universe. It is concerned with fundamental questions about its creation and evolution. In order to develop a better understanding of physical cosmology, it is important to understand big bang nucleosynthesis, absolute time and space, dark energy and Friedmann–Lemaître–Robertson–Walker metric. This chapter is an overview of the subject matter incorporating all the major aspects of physical cosmology.

Physical Cosmology

Physical cosmology is the study of the largest-scale structures and dynamics of the Universe and is concerned with fundamental questions about its origin, structure, evolution, and ultimate fate. Cosmology as a science originated with the Copernican principle, which implies that celestial bodies obey identical physical laws to those on Earth, and Newtonian mechanics, which first allowed us to understand those physical laws. Physical cosmology, as it is now understood, began with the development in 1915 of Albert Einstein's general theory of relativity, followed by major observational discoveries in the 1920s: first, Edwin Hubble discovered that the universe contains a huge number of external galaxies beyond our own Milky Way; then, work by Vesto Slipher and others showed that the universe is expanding. These advances made it possible to speculate about the origin of the universe, and allowed the establishment of the Big Bang Theory, by Georges Lemaitre, as the leading cosmological model. A few researchers still advocate a handful of alternative cosmologies; however, most cosmologists agree that the Big Bang theory explains the observations better.

Dramatic advances in observational cosmology since the 1990s, including the cosmic microwave background, distant supernovae and galaxy redshift surveys, have led to the development of a standard model of cosmology. This model requires the universe to contain large amounts of dark matter and dark energy whose nature is currently not well understood, but the model gives detailed predictions that are in excellent agreement with many diverse observations.

Cosmology draws heavily on the work of many disparate areas of research in theoretical and applied physics. Areas relevant to cosmology include particle physics experiments and theory, theoretical and observational astrophysics, general relativity, quantum mechanics, and plasma physics.

Subject History

Modern cosmology developed along tandem tracks of theory and observation. In 1916, Albert Einstein published his theory of general relativity, which provided a unified description of gravity as a geometric property of space and time. At the time, Einstein believed in a static universe, but found that his original formulation of the theory did not permit it. This is because masses distributed throughout the universe gravitationally attract, and move toward each other over time. However,

he realized that his equations permitted the introduction of a constant term which could counteract the attractive force of gravity on the cosmic scale. Einstein published his first paper on relativistic cosmology in 1917, in which he added this *cosmological constant* to his field equations in
order to force them to model a static universe. However, this so-called Einstein model is unstable
to small perturbations—it will eventually start to expand or contract. The Einstein model describes
a static universe; space is finite and unbounded (analogous to the surface of a sphere, which has
a finite area but no edges). It was later realized that Einstein's model was just one of a larger set
of possibilities, all of which were consistent with general relativity and the cosmological principle.
The cosmological solutions of general relativity were found by Alexander Friedmann in the early
1920s. His equations describe the Friedmann–Lemaître–Robertson–Walker universe, which may
expand or contract, and whose geometry may be open, flat, or closed.

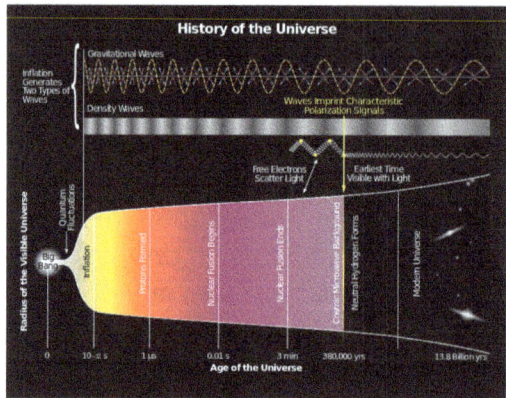

History of the Universe – gravitational waves are hypothesized to arise from cosmic inflation,
a faster-than-light expansion just after the Big Bang

In the 1910s, Vesto Slipher (and later Carl Wilhelm Wirtz) interpreted the red shift of spiral nebulae
as a Doppler shift that indicated they were receding from Earth. However, it is difficult to determine the distance to astronomical objects. One way is to compare the physical size of an object to
its angular size, but a physical size must be assumed to do this. Another method is to measure the
brightness of an object and assume an intrinsic luminosity, from which the distance may be determined using the inverse square law. Due to the difficulty of using these methods, they did not realize
that the nebulae were actually galaxies outside our own Milky Way, nor did they speculate about
the cosmological implications. In 1927, the Belgian Roman Catholic priest Georges Lemaître independently derived the Friedmann–Lemaître–Robertson–Walker equations and proposed, on the
basis of the recession of spiral nebulae, that the universe began with the "explosion" of a "primeval
atom"—which was later called the Big Bang. In 1929, Edwin Hubble provided an observational basis
for Lemaître's theory. Hubble showed that the spiral nebulae were galaxies by determining their
distances using measurements of the brightness of Cepheid variable stars. He discovered a relationship between the redshift of a galaxy and its distance. He interpreted this as evidence that the galaxies are receding from Earth in every direction at speeds proportional to their distance. This fact is
now known as Hubble's law, though the numerical factor Hubble found relating recessional velocity
and distance was off by a factor of ten, due to not knowing about the types of Cepheid variables.

Given the cosmological principle, Hubble's law suggested that the universe was expanding. Two
primary explanations were proposed for the expansion. One was Lemaître's Big Bang theory, advocated and developed by George Gamow. The other explanation was Fred Hoyle's steady state

model in which new matter is created as the galaxies move away from each other. In this model, the universe is roughly the same at any point in time.

For a number of years, support for these theories was evenly divided. However, the observational evidence began to support the idea that the universe evolved from a hot dense state. The discovery of the cosmic microwave background in 1965 lent strong support to the Big Bang model, and since the precise measurements of the cosmic microwave background by the Cosmic Background Explorer in the early 1990s, few cosmologists have seriously proposed other theories of the origin and evolution of the cosmos. One consequence of this is that in standard general relativity, the universe began with a singularity, as demonstrated by Roger Penrose and Stephen Hawking in the 1960s.

An alternative view to extend the Big Bang model, suggesting the universe had no beginning or singularity and the age of the universe is infinite, has been presented.

Energy of the Cosmos

Light chemical elements, primarily hydrogen and helium, were created in the Big Bang process. The small atomic nuclei combined into larger atomic nuclei to form heavier elements such as iron and nickel, which are more stable. This caused a *later energy release*. Such reactions of nuclear particles inside stars continue to contribute to *sudden energy releases*, such as in nova stars. Gravitational collapse of matter into black holes is also thought to power the most energetic processes, generally seen at the centers of galaxies.

Cosmologists cannot explain all cosmic phenomena exactly, such as those related to the accelerating expansion of the universe, using conventional forms of energy. Instead, cosmologists propose a new form of energy called dark energy that permeates all space. One hypothesis is that dark energy is the energy of virtual particles, which are believed to exist in a vacuum due to the uncertainty principle.

There is no clear way to define the total energy in the universe using the most widely accepted theory of gravity, general relativity. Therefore, it remains controversial whether the total energy is conserved in an expanding universe. For instance, each photon that travels through intergalactic space loses energy due to the redshift effect. This energy is not obviously transferred to any other system, so seems to be permanently lost. On the other hand, some cosmologists insist that energy is conserved in some sense; this follows the law of conservation of energy.

Thermodynamics of the universe is a field of study that explores which form of energy dominates the cosmos – relativistic particles which are referred to as radiation, or non-relativistic particles referred to as matter. Relativistic particles are particles whose rest mass is zero or negligible compared to their kinetic energy, and so move at the speed of light or very close to it; non-relativistic particles have much higher rest mass than their energy and so move much slower than the speed of light.

As the universe expands, both matter and radiation in it become diluted. However, the energy densities of radiation and matter dilute at different rates. As a particular volume expands, mass energy density is changed only by the increase in volume, but the energy density of radiation is changed both by the increase in volume and by the increase in the wavelength of the photons that make it up. Thus the energy of radiation becomes a smaller part of the universe's total energy than that of matter as it expands. The very early universe is said to have been 'radiation dominated' and radiation controlled the deceleration of expansion. Later, as the average energy per photon becomes roughly 10 eV and lower, matter dictates the rate of deceleration and the universe is said to be 'matter dominated'. The intermediate case is not treated well analytically. As the expansion of the universe continues, matter dilutes even further and the cosmological constant becomes dominant, leading to an acceleration in the universe's expansion.

History of the Universe

The history of the universe is a central issue in cosmology. The history of the universe is divided into different periods called epochs, according to the dominant forces and processes in each period. The standard cosmological model is known as the Lambda-CDM model.

Equations of Motion

The equations of motion governing the universe as a whole are derived from general relativity with a small, positive cosmological constant. The solution is an expanding universe; due to this expansion, the radiation and matter in the universe cool down and become diluted. At first, the expansion is slowed down by gravitation attracting the radiation and matter in the universe. However, as these become diluted, the cosmological constant becomes more dominant and the expansion of the universe starts to accelerate rather than decelerate. In our universe this happened billions of years ago.

Particle Physics in Cosmology

Particle physics is important to the behavior of the early universe, because the early universe was so hot that the average energy density was very high. Because of this, scattering processes and decay of unstable particles are important in cosmology.

As a rule of thumb, a scattering or a decay process is cosmologically important in a certain cosmological epoch if the time scale describing that process is smaller than, or comparable to, the time scale of the expansion of the universe. The time scale that describes the expansion of the universe is $1/H$ with H being the Hubble constant, which itself actually varies with time. The expansion timescale $1/H$ is roughly equal to the age of the universe at that time.

Timeline of the Big Bang

Observations suggest that the universe began around 13.8 billion years ago. Since then, the evolution of the universe has passed through three phases. The very early universe, which is still poorly understood, was the split second in which the universe was so hot that particles had energies higher than those currently accessible in particle accelerators on Earth. Therefore, while the basic features of this epoch have been worked out in the Big Bang theory, the details are largely based on educated guesses. Following this, in the early universe, the evolution of the universe proceeded according to known high energy physics. This is when the first protons, electrons and neutrons formed, then nuclei and finally atoms. With the formation of neutral hydrogen, the cosmic microwave background was emitted. Finally, the epoch of structure formation began, when matter started to aggregate into the first stars and quasars, and ultimately galaxies, clusters of galaxies and superclusters formed. The future of the universe is not yet firmly known, but according to the ΛCDM model it will continue expanding forever.

Areas of Study

Below, some of the most active areas of inquiry in cosmology are described, in roughly chronological order. This does not include all of the Big Bang cosmology, which is presented in *Timeline of the Big Bang*.

Very Early Universe

The early, hot universe appears to be well explained by the Big Bang from roughly 10^{-33} seconds onwards, but there are several problems. One is that there is no compelling reason, using current particle physics, for the universe to be flat, homogeneous, and isotropic. Moreover, grand unified theories of particle physics suggest that there should be magnetic monopoles in the universe, which have not been found. These problems are resolved by a brief period of cosmic inflation, which drives the universe to flatness, smooths out anisotropies and inhomogeneities to the observed level, and exponentially dilutes the monopoles. The physical model behind cosmic inflation is extremely simple, but it has not yet been confirmed by particle physics, and there are difficult problems reconciling inflation and quantum field theory. Some cosmologists think that string theory and brane cosmology will provide an alternative to inflation.

Another major problem in cosmology is what caused the universe to contain far more matter than antimatter. Cosmologists can observationally deduce that the universe is not split into regions of matter and antimatter. If it were, there would be X-rays and gamma rays produced as a result of annihilation, but this is not observed. Therefore, some process in the early universe must have created a small excess of matter over antimatter, and this (currently not understood) process is called *baryogenesis*. Three required conditions for baryogenesis were derived by Andrei Sakharov in 1967, and requires a violation of the particle physics symmetry, called CP-symmetry, between matter and antimatter. However, particle accelerators measure too small a violation of CP-symmetry to account for the baryon asymmetry. Cosmologists and particle physicists look for additional violations of the CP-symmetry in the early universe that might account for the baryon asymmetry.

Both the problems of baryogenesis and cosmic inflation are very closely related to particle physics, and their resolution might come from high energy theory and experiment, rather than through observations of the universe.

Big Bang Theory

Big Bang nucleosynthesis is the theory of the formation of the elements in the early universe. It finished when the universe was about three minutes old and its temperature dropped below that at which nuclear fusion could occur. Big Bang nucleosynthesis had a brief period during which it could operate, so only the very lightest elements were produced. Starting from hydrogen ions (protons), it principally produced deuterium, helium-4, and lithium. Other elements were produced in only trace abundances. The basic theory of nucleosynthesis was developed in 1948 by George Gamow, Ralph Asher Alpher, and Robert Herman. It was used for many years as a probe of physics at the time of the Big Bang, as the theory of Big Bang nucleosynthesis connects the abundances of primordial light elements with the features of the early universe. Specifically, it can be used to test the equivalence principle, to probe dark matter, and test neutrino physics. Some cosmologists have proposed that Big Bang nucleosynthesis suggests there is a fourth "sterile" species of neutrino.

Standard Model of Big Bang Cosmology

The ΛCDM (Lambda cold dark matter) or Lambda-CDM model is a parametrization of the Big Bang cosmological model in which the universe contains a cosmological constant, denoted by Lambda (Greek Λ), associated with dark energy, and cold dark matter (abbreviated CDM). It is frequently referred to as the standard model of Big Bang cosmology.

Cosmic Microwave Background

The cosmic microwave background is radiation left over from decoupling after the epoch of recombination when neutral atoms first formed. At this point, radiation produced in the Big Bang stopped Thomson scattering from charged ions. The radiation, first observed in 1965 by Arno Penzias and Robert Woodrow Wilson, has a perfect thermal black-body spectrum. It has a temperature of 2.7 kelvins today and is isotropic to one part in 10^5. Cosmological perturbation theory, which describes the evolution of slight inhomogeneities in the early universe, has allowed cosmologists to precisely calculate the angular power spectrum of the radiation, and it has been measured by the recent satellite experiments (COBE and WMAP) and many ground and balloon-based experiments (such as Degree Angular Scale Interferometer, Cosmic Background Imager, and Boomerang). One of the goals of these efforts is to measure the basic parameters of the Lambda-CDM model with increasing accuracy, as well as to test the predictions of the Big Bang model and look for new physics. The recent measurements made by WMAP, for example, have placed limits on the neutrino masses.

Newer experiments, such as QUIET and the Atacama Cosmology Telescope, are trying to measure the polarization of the cosmic microwave background. These measurements are expected to provide further confirmation of the theory as well as information about cosmic inflation, and the so-called secondary anisotropies, such as the Sunyaev-Zel'dovich effect and Sachs-Wolfe effect, which are caused by interaction between galaxies and clusters with the cosmic microwave background.

On 17 March 2014, astronomers of the BICEP2 Collaboration announced the apparent detection of B-mode polarization of the CMB, considered to be evidence of primordial gravitational waves that are predicted by the theory of inflation to occur during the earliest phase of the Big Bang. However, later that year the Planck collaboration provided a more accurate measurement of cosmic dust, concluding that the B-mode signal from dust is the same strength as that reported from BICEP2. On January 30, 2015, a joint analysis of BICEP2 and Planck data was published and the European Space Agency announced that the signal can be entirely attributed to interstellar dust in the Milky Way.

Formation and Evolution of Large-scale Structure

Understanding the formation and evolution of the largest and earliest structures (i.e., quasars, galaxies, clusters and superclusters) is one of the largest efforts in cosmology. Cosmologists study a model of hierarchical structure formation in which structures form from the bottom up, with smaller objects forming first, while the largest objects, such as superclusters, are still assembling. One way to study structure in the universe is to survey the visible galaxies, in order to construct a three-dimensional picture of the galaxies in the universe and measure the matter power spectrum. This is the approach of the *Sloan Digital Sky Survey* and the 2dF Galaxy Redshift Survey.

Another tool for understanding structure formation is simulations, which cosmologists use to study the gravitational aggregation of matter in the universe, as it clusters into filaments, superclusters and voids. Most simulations contain only non-baryonic cold dark matter, which should suffice to understand the universe on the largest scales, as there is much more dark matter in the universe than visible, baryonic matter. More advanced simulations are starting to include baryons and study the formation of individual galaxies. Cosmologists study these simulations to see if they agree with the galaxy surveys, and to understand any discrepancy.

Other, complementary observations to measure the distribution of matter in the distant universe and to probe reionization include:

- The Lyman-alpha forest, which allows cosmologists to measure the distribution of neutral atomic hydrogen gas in the early universe, by measuring the absorption of light from distant quasars by the gas.

- The 21 centimeter absorption line of neutral atomic hydrogen also provides a sensitive test of cosmology

- Weak lensing, the distortion of a distant image by gravitational lensing due to dark matter.

These will help cosmologists settle the question of when and how structure formed in the universe.

Dark Matter

Evidence from Big Bang nucleosynthesis, the cosmic microwave background and structure formation suggests that about 23% of the mass of the universe consists of non-baryonic dark matter, whereas only 4% consists of visible, baryonic matter. The gravitational effects of dark matter are well understood, as it behaves like a cold, non-radiative fluid that forms haloes around galaxies. Dark matter has never been detected in the laboratory, and the particle physics nature of dark matter remains completely unknown. Without observational constraints, there are a number of candidates, such as a stable supersymmetric particle, a weakly interacting massive particle, an axion, and a massive compact halo object. Alternatives to the dark matter hypothesis include a modification of gravity at small accelerations (MOND) or an effect from brane cosmology.

Dark Energy

If the universe is flat, there must be an additional component making up 73% (in addition to the 23% dark matter and 4% baryons) of the energy density of the universe. This is called dark energy. In order not to interfere with Big Bang nucleosynthesis and the cosmic microwave background, it must not cluster in haloes like baryons and dark matter. There is strong observational evidence for dark energy, as the total energy density of the universe is known through constraints on the flatness of the universe, but the amount of clustering matter is tightly measured, and is much less than this. The case for dark energy was strengthened in 1999, when measurements demonstrated that the expansion of the universe has begun to gradually accelerate.

Apart from its density and its clustering properties, nothing is known about dark energy. *Quantum field theory* predicts a cosmological constant (CC) much like dark energy, but 120 orders of magnitude larger than that observed. Steven Weinberg and a number of string theorists have invoked the 'weak anthropic principle': i.e. the reason that physicists observe a universe with such a small cosmological constant is that no physicists (or any life) could exist in a universe with a larger cosmological constant. Many cosmologists find this an unsatisfying explanation: perhaps because while the weak anthropic principle is self-evident (given that living observers exist, there must be at least one universe with a cosmological constant which allows for life to exist) it does not attempt to explain the context of that universe. For example, the weak anthropic principle alone does not distinguish between:

- Only one universe will ever exist and there is some underlying principle that constrains the CC to the value we observe.

- Only one universe will ever exist and although there is no underlying principle fixing the CC, we got lucky.

- Lots of universes exist (simultaneously or serially) with a range of CC values, and of course ours is one of the life-supporting ones.

Other possible explanations for dark energy include quintessence or a modification of gravity on the largest scales. The effect on cosmology of the dark energy that these models describe is given by the dark energy's equation of state, which varies depending upon the theory. The nature of dark energy is one of the most challenging problems in cosmology.

A better understanding of dark energy is likely to solve the problem of the ultimate fate of the universe. In the current cosmological epoch, the accelerated expansion due to dark energy is preventing structures larger than superclusters from forming. It is not known whether the acceleration will continue indefinitely, perhaps even increasing until a big rip, or whether it will eventually reverse.

Gravitational Waves

Gravitational waves are ripples in the curvature of spacetime that propagate as waves at the speed of light, generated in certain gravitational interactions that propagate outward from their source. Gravitational-wave astronomy is an emerging branch of observational astronomy which aims to use gravitational waves to collect observational data about sources of detectable gravitational waves such as binary star systems composed of white dwarfs, neutron stars, and black holes; and events such as supernovae, and the formation of the early universe shortly after the Big Bang.

In 2016, the LIGO Scientific Collaboration and Virgo Collaboration teams announced that they had made the first observation of gravitational waves, originating from a pair of merging black holes using the Advanced LIGO detectors. On June 15, 2016, a second detection of gravitational waves from coalescing black holes was announced. Besides LIGO, many other gravitational-wave observatories (detectors) are under construction.

Other Areas of Inquiry

Cosmologists also study:

- Whether primordial black holes were formed in our universe, and what happened to them.

- The GZK cutoff for high-energy cosmic rays, and whether it signals a failure of special relativity at high energies

- The equivalence principle, whether or not Einstein's general theory of relativity is the correct theory of gravitation, and if the fundamental laws of physics are the same everywhere in the universe.

- The increasing complexity of universal structures, an example being the progressively greater energy rate density.

Thermodynamics of the Universe

The thermodynamics of the universe is dictated by which form of energy dominates it - relativistic particles which are referred to as radiation, or non-relativistic particles which are referred to as matter. The former are particles whose rest mass is zero or negligible compared to their energy, and therefore move at the speed of light or very close to it; the latter are particles whose kinetic energy is much lower than their rest mass and therefore move much slower than the speed of light. The intermediate case is not treated well analytically.

Energy Density in the Expanding Universe

If the universe is expanding adiabatically then it will satisfy the first law of thermodynamics:

$$0 = dQ = dU + PdV$$

where Q is the total heat which is assumed to be constant, U is the internal energy of the matter and radiation in the universe, P is the pressure and V the volume.

One then finds an equation for the energy density $u \equiv U/V$, and so

$$du = d\left(\frac{U}{V}\right) = \frac{dU}{V} - U\frac{dV}{V^2} = -(p+u)\frac{dV}{V} = -3(p+u)\frac{da}{a}$$

where in the last equality we used the fact that the total volume of the universe is proportional to a^3, a being the scale factor of the universe.

In fact this equation can be directly obtained from the equations of motion governing the Friedmann-Lemaître-Robertson-Walker metric: by dividing the equation above with dt and identifying $\rho = u$ (the energy density), we get one of the FLRW equations of motions.

In the comoving coordinates, u is equal to the mass density ρ. For radiation, $p = u/3$ whereas for matter $p \ll u$ and the pressure can be neglected. Thus we get:

For radiation $du = -4u \dfrac{da}{a}$ thus u is proportional to a^{-4}

For matter $du = -3u \dfrac{da}{a}$ thus u is proportional to a^{-3}

This can be understood as follows: For matter, the energy density is equal (in our approximation) to the rest mass density. This is inversely proportional to the volume, and is therefore proportional to a^{-3}. For radiation, the energy density depends on the temperature T as well, and is therefore proportional to Ta^{-3}. As the universe expands it cools down, so T depends on a as well. In fact, since the energy of a relativistic particle is inversely proportional to its wavelength, which is proportional to a, the energy density of the radiation must be proportional to a^{-4}.

From this discussion it is also obvious that the temperature of radiation is inversely proportional to the scale factor a.

Rate of Expansion of the Universe

Plugging this information to the Friedmann-Lemaître-Robertson-Walker equations of motion and neglecting both the cosmological constant Λ and the curvature parameter k, which is justified for the early universe ($a \ll 1$), one gets the following equation:

$$\dot{a}^2 \propto a^2 \rho$$

$\rho = u$ is the energy density, and one finds the following behavior:

- In a radiation-dominated universe: $a \propto t^{1/2}$

- In a matter-dominated universe: $a \propto t^{2/3}$

One can further show that the universe was radiation-dominated as long as the energy density was of the order of 10 eV to the fourth, or higher. Since the energy density keeps going down, this was no longer true when the universe was 70,000 years old, when it became matter dominant.

In the universe today, matter is mainly in forms of galaxies and dark matter, while the radiation is the cosmic microwave background radiation, the cosmic neutrino background (if the neutrino rest mass is high enough then the latter is formally matter), and finally, mostly in the form of dark energy.

Dark Energy and Cosmic Inflation

Dark energy is a hypothetical form of energy that permeates all of space and its negative pressure coincides with an acceleration in the expansion of the universe. Positive pressure coincides with a deceleration as does the gravity of energy and mass. There is no known cause and effect in fundamental physics, so it is not assumed the pressures or gravity "cause" a reduction or acceleration in the expansion of the universe, nor vice versa. For example, the energy in the gravitional field of the universe that coincides with its expansion is equal and opposite to the mass energy of the universe and it is not assumed (and the equations do not indicate) that the expansion created the positive mass energy and negative gravitational energy, nor vice versa.

According to the equation above,

$$\dot{u} = -3(p+u)\frac{\dot{a}}{a}$$

Thus the more negative the pressure is, the less the energy density reduces as the universe expands. In other words, Dark energy dilutes less than any other form of energy, and will therefore eventually dominate the universe, as all other energy densities gets diluted faster with the expansion of the universe.

In fact, if the dark energy is created by a cosmological constant or a constant scalar field, then its pressure is minus its energy density $p = -u$, and therefore its energy density remains constant (as is expected by definition).

Dark energy is usually assumed to be the Casimir energy of the vacuum, with possible contributions from the energy density of scalar fields which has a non-zero value at the vacuum. It may be that this field can decay at some time in the distant future, leading to a new vacuum state, different than the one we are living in. This is a phase transition, where the dark energy is reduced and huge amounts of energy in conventional forms (i.e. particles) are produced.

Such a series of events is in fact thought to have already occurred in the early universe, where first a cosmological constant much larger than the present one came to dominate the universe, bringing about cosmic inflation. At the end of this epoch, a phase transition occurred where the cosmological constant was reduced to its present value and huge amounts of energy where produced, from which all the radiation and matter of the early universe came about.

Particle Physics in Cosmology

Particle physics, which deals with the interactions of elementary particles at high energies, is an important component of cosmological models of the early universe, when the universe was dominated by radiation and its average energy density was very high. Because of this, pair production, scattering processes and decay of unstable particles are important in cosmology, and the interface between particle physics and cosmology is sometimes referred to as particle cosmology.

As a thumb rule, a scattering or a decay process is cosmologically important in a certain cosmological epoch if its relevant time scale is smaller or even to the time scale of the universe expansion, which is $\frac{1}{H}$ with H being the Hubble constant at that time. This is roughly equal to the age of the universe at that time.

For example, the pion has a lifetime of about 26 nanoseconds. This means that particle physics processes involving pions did not take place until roughly that much time passed since the start of the Big Bang.

Cosmological observations of phenomena such as the cosmic microwave background and the cosmic abundance of elements, together with the predictions of the Standard Model of particle physics, place constraints on the conditions of the early universe. The success of the Standard

Model at explaining these observations provides a confirmation of its validity outside of laboratory conditions. In addition, phenomena extrapolated from cosmological observations, such as dark matter and CP-violation, suggest a need for physics that goes beyond the Standard Model.

Friedmann–Lemaître–Robertson–Walker Metric

The Friedmann–Lemaître–Robertson–Walker (FLRW) metric is an exact solution of Einstein's field equations of general relativity; it describes a homogeneous, isotropic expanding or contracting universe that is path connected, but not necessarily simply connected. The general form of the metric follows from the geometric properties of homogeneity and isotropy; Einstein's field equations are only needed to derive the scale factor of the universe as a function of time. Depending on geographical or historical preferences, the set of the four scientists — Alexander Friedmann, Georges Lemaître, Howard P. Robertson and Arthur Geoffrey Walker are customarily grouped as Friedmann–Robertson–Walker (FRW) or Robertson–Walker (RW) or Friedmann–Lemaître (FL)). This model is sometimes called the *Standard Model* of modern cosmology, although such a description is also associated with the further developed Lambda-CDM model. The FLRW model was developed independently by the named authors in the 1920s and 1930s.

General Metric

The FLRW metric starts with the assumption of homogeneity and isotropy of space. It also assumes that the spatial component of the metric can be time-dependent. The generic metric which meets these conditions is

$$-c^2 d\tau^2 = -c^2 dt^2 + a(t)^2 d\Sigma^2$$

where Σ ranges over a 3-dimensional space of uniform curvature, that is, elliptical space, Euclidean space, or hyperbolic space. It is normally written as a function of three spatial coordinates, but there are several conventions for doing so, detailed below. $d\Sigma$ does not depend on t — all of the time dependence is in the function $a(t)$, known as the "scale factor".

Reduced-circumference Polar Coordinates

In reduced-circumference polar coordinates the spatial metric has the form

$$d\Sigma^2 = \frac{dr^2}{1 - kr^2} + r^2 d\Omega^2, \quad \text{where } d\Omega^2 = d\theta^2 + \sin^2\theta d\phi^2.$$

k is a constant representing the curvature of the space. There are two common unit conventions:

- k may be taken to have units of length^{-2}, in which case r has units of length and $a(t)$ is unitless. k is then the Gaussian curvature of the space at the time when $a(t) = 1$. r is sometimes called the reduced circumference because it is equal to the measured circumference of a circle (at that value of r), centered at the origin, divided by 2π (like the r of Schwarzschild coordinates). Where appropriate, $a(t)$ is often chosen to equal 1 in the present cosmological era, so that $d\Sigma$ measures comoving distance.

- Alternatively, k may be taken to belong to the set $\{-1,0,+1\}$ (for negative, zero, and positive curvature respectively). Then r is unitless and $a(t)$ has units of length. When $k = \pm 1$, $a(t)$ is the radius of curvature of the space, and may also be written $R(t)$.

A disadvantage of reduced circumference coordinates is that they cover only half of the 3-sphere in the case of positive curvature—circumferences beyond that point begin to decrease, leading to degeneracy. (This is not a problem if space is elliptical, i.e. a 3-sphere with opposite points identified.)

Hyperspherical Coordinates

In *hyperspherical* or *curvature-normalized* coordinates the coordinate r is proportional to radial distance; this gives

$$d\Sigma^2 = dr^2 + S_k(r)^2 \, d\Omega^2$$

where $d\Omega$ is as before and

$$S_k(r) = \begin{cases} \sqrt{k}^{-1} \sin(r\sqrt{k}), & k > 0 \\ r, & k = 0 \\ \sqrt{|k|}^{-1} \sinh(r\sqrt{|k|}), & k < 0. \end{cases}$$

As before, there are two common unit conventions:

- k may be taken to have units of length^{-2}, in which case r has units of length and $a(t)$ is unitless. k is then the Gaussian curvature of the space at the time when $a(t) = 1$. Where appropriate, $a(t)$ is often chosen to equal 1 in the present cosmological era, so that $d\Sigma$ measures comoving distance.

- Alternatively, as before, k may be taken to belong to the set $\{-1,0,+1\}$ (for negative, zero, and positive curvature respectively). Then r is unitless and $a(t)$ has units of length. When $k = \pm 1$, $a(t)$ is the radius of curvature of the space, and may also be written $R(t)$. Note that, when $k = +1$, r is essentially a third angle along with θ and φ. The letter χ may be used instead of r.

Though it is usually defined piecewise as above, S is an analytic function of both k and r. It can also be written as a power series

$$S_k(r) = \sum_{n=0}^{\infty} \frac{(-1)^n k^n r^{2n+1}}{(2n+1)!} = r - \frac{kr^3}{6} + \frac{k^2 r^5}{120} - \cdots$$

or as

$$S_k(r) = r \, \mathrm{sinc}(r\sqrt{k})$$

where sinc is the unnormalized sinc function and $\sqrt{k}$ is one of the imaginary, zero or real square roots of k. These definitions are valid for all k.

Cartesian Coordinates

When $k = 0$ one may write simply

$$d\Sigma^2 = dx^2 + dy^2 + dz^2.$$

This can be extended to $k \neq 0$ by defining

$$x = r \cos \theta,$$

$$y = r \sin \theta \cos \phi \text{, and}$$

$$z = r \sin \theta \sin \phi,$$

where r is one of the radial coordinates defined above, but this is rare.

Curvature

In flat (k=0) FRW space using Cartesian coordinates, the surviving components of the Ricci tensor are

$$R_{tt} = -3\frac{\ddot{a}}{a}, \quad R_{xx} = R_{yy} = R_{zz} = c^{-2}(a\ddot{a} + 2\dot{a}^2)$$

and the Ricci scalar is

$$R = 6c^{-2}\left(\frac{\ddot{a}(t)}{a(t)} + \frac{\dot{a}^2(t)}{a^2(t)}\right).$$

Spherical Coordinates

In more general FRW space using spherical coordinates (called "reduced-circumference polar co-ordinates" above), the surviving components of the Ricci tensor are

$$R_{tt} = -3\frac{\ddot{a}}{a},$$

$$R_{rr} = \frac{c^{-2}(a(t)\ddot{a}(t) + 2\dot{a}^2(t)) + 2k}{1 - kr^2}$$

$$R_{\theta\theta} = r^2(c^{-2}(a(t)\ddot{a}(t) + 2\dot{a}^2(t)) + 2k)$$

$$R_{\phi\phi} = r^2(c^{-2}(a(t)\ddot{a}(t) + 2\dot{a}^2(t)) + 2k)\sin^2(\theta)$$

and the Ricci scalar is

$$R = 6\left(\frac{\ddot{a}(t)}{c^2 a(t)} + \frac{\dot{a}^2(t)}{c^2 a^2(t)} + \frac{k}{a^2(t)} \right).$$

Solutions

Einstein's field equations are not used in deriving the general form for the metric: it follows from the geometric properties of homogeneity and isotropy. However, determining the time evolution of $a(t)$ does require Einstein's field equations together with a way of calculating the density, $\rho(t)$, such as a cosmological equation of state.

This metric has an analytic solution to Einstein's field equations $G_{\mu\nu} + \Lambda g_{\mu\nu} = \frac{8\pi G}{c^4} T_{\mu\nu}$ giving the Friedmann equations when the energy-momentum tensor is similarly assumed to be isotropic and homogeneous. The resulting equations are:

$$\left(\frac{\dot{a}}{a} \right)^2 + \frac{kc^2}{a^2} - \frac{\Lambda c^2}{3} = \frac{8\pi G}{3} \rho$$

$$2\frac{\ddot{a}}{a} + \left(\frac{\dot{a}}{a} \right)^2 + \frac{kc^2}{a^2} - \Lambda c^2 = -\frac{8\pi G}{c^2} p.$$

These equations are the basis of the standard big bang cosmological model including the current ΛCDM model. Because the FLRW model assumes homogeneity, some popular accounts mistakenly assert that the big bang model cannot account for the observed lumpiness of the universe. In a strictly FLRW model, there are no clusters of galaxies, stars or people, since these are objects much denser than a typical part of the universe. Nonetheless, the FLRW model is used as a first approximation for the evolution of the real, lumpy universe because it is simple to calculate, and models which calculate the lumpiness in the universe are added onto the FLRW models as extensions. Most cosmologists agree that the observable universe is well approximated by an *almost FLRW model*, i.e., a model which follows the FLRW metric apart from primordial density fluctuations. As of 2003, the theoretical implications of the various extensions to the FLRW model appear to be well understood, and the goal is to make these consistent with observations from COBE and WMAP.

If the spacetime is multiply connected, then each event in spacetime will be represented by more than one tuple of coordinates.

Interpretation

The pair of equations given above is equivalent to the following pair of equations

$$\dot{\rho} = -3\frac{\dot{a}}{a}\left(\rho + \frac{p}{c^2} \right)$$

$$\frac{\ddot{a}}{a} = -\frac{4\pi G}{3}\left(\rho + \frac{3p}{c^2}\right) + \frac{\Lambda c^2}{3}$$

with k, the spatial curvature index, serving as a constant of integration for the first equation.

The first equation can be derived also from thermodynamical considerations and is equivalent to the first law of thermodynamics, assuming the expansion of the universe is an adiabatic process (which is implicitly assumed in the derivation of the Friedmann–Lemaître–Robertson–Walker metric).

The second equation states that both the energy density and the pressure cause the expansion rate of the universe $\dot{a}$ to decrease, i.e., both cause a deceleration in the expansion of the universe. This is a consequence of gravitation, with pressure playing a similar role to that of energy (or mass) density, according to the principles of general relativity. The cosmological constant, on the other hand, causes an acceleration in the expansion of the universe.

Cosmological Constant

The cosmological constant term can be omitted if we make the following replacements

$$\rho \to \rho + \frac{\Lambda c^2}{8\pi G}$$

$$p \to p - \frac{\Lambda c^4}{8\pi G}.$$

Therefore, the cosmological constant can be interpreted as arising from a form of energy which has negative pressure, equal in magnitude to its (positive) energy density:

$$p = -\rho c^2.$$

Such form of energy—a generalization of the notion of a cosmological constant—is known as dark energy.

In fact, in order to get a term which causes an acceleration of the universe expansion, it is enough to have a scalar field which satisfies

$$p < -\frac{\rho c^2}{3}.$$

Such a field is sometimes called quintessence.

Newtonian Interpretation

This is due to McCrea and Milne although sometimes incorrectly ascribed to Friedmann. The Friedmann equations are equivalent to this pair of equations:

$$-a^3\dot{\rho} = 3a^2\dot{a}\rho + \frac{3a^2 p\dot{a}}{c^2}$$

$$\frac{\dot{a}^2}{2} - \frac{G\dfrac{4\pi a^3}{3}\rho}{a} = -\frac{kc^2}{2}$$

The first equation says that the decrease in the mass contained in a fixed cube (whose side is momentarily a) is the amount which leaves through the sides due to the expansion of the universe plus the mass equivalent of the work done by pressure against the material being expelled. This is the conservation of mass-energy (first law of thermodynamics) contained within a part of the universe.

The second equation says that the kinetic energy (seen from the origin) of a particle of unit mass moving with the expansion plus its (negative) gravitational potential energy (relative to the mass contained in the sphere of matter closer to the origin) is equal to a constant related to the curvature of the universe. In other words, the energy (relative to the origin) of a co-moving particle in free-fall is conserved. General relativity merely adds a connection between the spatial curvature of the universe and the energy of such a particle: positive total energy implies negative curvature and negative total energy implies positive curvature.

The cosmological constant term is assumed to be treated as dark energy and thus merged into the density and pressure terms.

During the Planck epoch, one cannot neglect quantum effects. So they may cause a deviation from the Friedmann equations.

Name and History

The main results of the FLRW model were first derived by the Soviet mathematician Alexander Friedmann in 1922 and 1924. Although his work was published in the prestigious physics journal Zeitschrift für Physik, it remained relatively unnoticed by his contemporaries. Friedmann was in direct communication with Albert Einstein, who, on behalf of Zeitschrift für Physik, acted as the scientific referee of Friedmann's work. Eventually Einstein acknowledged the correctness of Friedmann's calculations, but failed to appreciate the physical significance of Friedmann's predictions.

Friedmann died in 1925. In 1927, Georges Lemaître, a Belgian priest, astronomer and periodic professor of physics at the Catholic University of Leuven, arrived independently at similar results as Friedmann had and published them in Annals of the Scientific Society of Brussels. In the face of the observational evidence for the expansion of the universe obtained by Edwin Hubble in the late 1920s, Lemaître's results were noticed in particular by Arthur Eddington, and in 1930–31 his paper was translated into English and published in the Monthly Notices of the Royal Astronomical Society.

Howard P. Robertson from the US and Arthur Geoffrey Walker from the UK explored the problem further during the 1930s. In 1935 Robertson and Walker rigorously proved that the FLRW metric is the only one on a spacetime that is spatially homogeneous and isotropic (as noted above, this is a geometric result and is not tied specifically to the equations of general relativity, which were always assumed by Friedmann and Lemaître).

Because the dynamics of the FLRW model were derived by Friedmann and Lemaître, the latter two names are often omitted by scientists outside the US. Conversely, US physicists often refer to it as simply "Robertson–Walker". The full four-name title is the most democratic and it is frequently used. Often the "Robertson–Walker" *metric*, so-called since they proved its generic properties, is distinguished from the dynamical "Friedmann-Lemaître" *models*, specific solutions for $a(t)$ which assume that the only contributions to stress-energy are cold matter ("dust"), radiation, and a cosmological constant.

Einstein's Radius of the Universe

Einstein's radius of the Universe is the radius of curvature of space of Einstein's universe, a long-abandoned static model that was supposed to represent our universe in idealized form. Putting

$$\dot{a} = \ddot{a} = 0.$$

in the Friedmann equation, the radius of curvature of space of this universe (Einstein's radius) is

$$R_E = c / \sqrt{4\pi G\rho},$$

where c is the speed of light, G is the Newtonian gravitational constant, and ρ is the density of space of this universe. The numerical value of Einstein's radius is of the order of 10^{10} light years.

Evidence

By combining the observation data from some experiments such as WMAP and Planck with theoretical results of Ehlers–Geren–Sachs theorem and its generalization, astrophysicists now agree that the universe is almost homogeneous and isotropic (when averaged over a very large scale) and thus nearly a FLRW spacetime.

Big Bang Nucleosynthesis

In physical cosmology, Big Bang nucleosynthesis (abbreviated BBN, also known as primordial nucleosynthesis, arch(a)eonucleosynthesis, archonucleosynthesis, protonucleosynthesis and pal(a) eonucleosynthesis) refers to the production of nuclei other than those of the lightest isotope of hydrogen (hydrogen-1, ^1H, having a single proton as a nucleus) during the early phases of the Universe. Primordial nucleosynthesis is believed by most cosmologists to have taken place in the interval from roughly 10 seconds to 20 minutes after the Big Bang, and is calculated to be responsible for the formation of most of the universe's helium as the isotope helium-4 (^4He), along with small amounts of the hydrogen isotope deuterium (^2H or D), the helium isotope helium-3 (^3He), and a very small amount of the lithium isotope lithium-7 (^7Li). In addition to these stable nuclei, two unstable or radioactive isotopes were also produced: the heavy hydrogen isotope tritium (^3H or T); and the beryllium isotope beryllium-7 (^7Be); but these unstable isotopes later decayed into ^3He and ^7Li, as above.

Essentially all of the elements that are heavier than lithium were created much later, by stellar nucleosynthesis in evolving and exploding stars.

Characteristics

There are several important characteristics of Big Bang nucleosynthesis (BBN):

- The initial conditions (neutron-proton ratio) were set in the first second after the Big Bang.

- The fusion of nuclei occurred between roughly 10 seconds to 20 minutes after the Big Bang; this corresponds to the temperature range when the universe was cool enough for deuterium to survive, but hot and dense enough for fusion reactions to occur at a significant rate.

- It was widespread, encompassing the entire observable universe.

The key parameter which allows one to calculate the effects of BBN is the baryon/photon number ratio, which is a small number of order 6×10^{-10}. This parameter corresponds to the baryon density and controls the rate at which nucleons collide and react; from this it is possible to calculate element abundances after nucleosynthesis ends Although the baryon per photon ratio is important in determining element abundances, the precise value makes little difference to the overall picture. Without major changes to the Big Bang theory itself, BBN will result in mass abundances of about 75% of hydrogen-1, about 25% helium-4, about 0.01% of deuterium and helium-3, trace amounts (on the order of 10^{-10}) of lithium, and negligible heavier elements. That the observed abundances in the universe are generally consistent with these abundance numbers is considered strong evidence for the Big Bang theory.

In this field, for historical reasons it is customary to quote the helium-4 fraction *by mass*, symbol Y, so that 25% helium-4 means that helium-4 atoms account for 25% of the mass, but less than 8% of the nuclei would be helium-4 nuclei. Other (trace) nuclei are usually expressed as number ratios to hydrogen.

Important Parameters

The creation of light elements during BBN was dependent on a number of parameters; among those was the neutron-proton ratio (calculable from Standard Model physics) and the baryon-photon ratio.

Neutron–proton Ratio

The neutron-proton ratio was set by Standard Model physics before the nucleosynthesis era, essentially within the first 1-second after the Big Bang. Neutrons can react with positrons or electron neutrinos to create protons and other products in one of the following reactions:

$$n + e^+ \leftrightarrow \text{anti-}\nu_e + p$$

$$n + \nu_e \leftrightarrow p + e^-$$

At times much earlier than 1 sec, these reactions were fast and maintained the n/p ratio close to 1:1. As the temperature dropped, the equilibrium shifted in favour of protons due to their slightly lower mass, and the n/p ratio smoothly decreased. These reactions continued until the decreasing temperature and density caused the reactions to become too slow, which occurred at about T = 0.7 MeV (time around 1 second) and is called the freeze out temperature. At freeze out, the neutron-proton ratio was about 1/6. However, free neutrons are unstable with a mean life of 880 sec; some neutrons decayed in the next few minutes before fusing into any nucleus, so the ratio of total neutrons to protons after nucleosynthesis ends is about 1/7. Almost all neutrons that fused instead of decaying ended up combined into helium-4, due to the fact that helium-4 has the highest binding energy per nucleon among light elements. This predicts that about 8% of all atoms should be helium-4, leading to a mass fraction of helium-4 of about 25%, which is in line with observations. Small traces of deuterium and helium-3 remained as there was insufficient time and density for them to react and form helium-4.

Baryon–photon Ratio

The baryon–photon ratio, η, is the key parameter determining the abundances of light elements after nucleosynthesis ends. Baryons and light elements can fuse in the following main reactions:

$p + n \rightarrow {}^{2}H + \gamma$

$p + {}^{2}H \rightarrow {}^{3}He + \gamma$

${}^{2}H + {}^{2}H \rightarrow {}^{3}He + n$

${}^{2}H + {}^{2}H \rightarrow {}^{3}H + p$

${}^{3}He + {}^{2}H \rightarrow {}^{4}He + p$

${}^{3}H + {}^{2}H \rightarrow {}^{4}He + n$

along with some other low-probability reactions leading to ${}^{7}Li$ or ${}^{7}Be$. (An important feature is that there are no stable nuclei with mass 5 or 8, which implies that reactions adding one baryon to ${}^{4}He$, or fusing two ${}^{4}He$, do not occur). Most fusion chains during BBN ultimately terminate in ${}^{4}He$ (helium-4), while "incomplete" reaction chains lead to small amounts of left-over ${}^{2}H$ or ${}^{3}He$; the amount of these decreases with increasing baryon-photon ratio. That is, the larger the baryon-photon ratio the more reactions there will be and the more efficiently deuterium will be eventually transformed into helium-4. This result makes deuterium a very useful tool in measuring the baryon-to-photon ratio.

Sequence

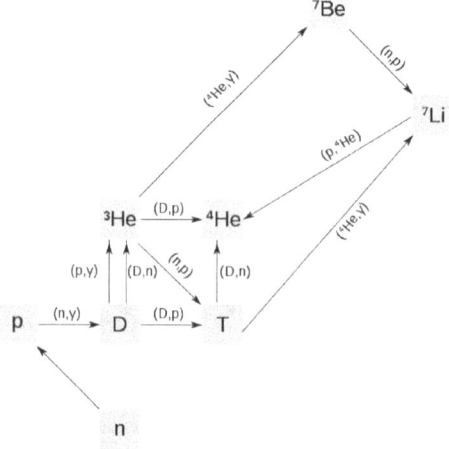

The main nuclear reaction chains for Big Bang nucleosynthesis

Big Bang nucleosynthesis began roughly 10 seconds after the big bang, when the universe had cooled sufficiently to allow deuterium nuclei to survive disruption by high-energy photons. (Note that the neutron-proton freeze-out time was earlier). This time is essentially independent of dark matter content, since the universe was highly radiation dominated until much later, and this dominant component controls the temperature/time relation. At this time there were about six protons for every neutron, but a small fraction of the neutrons decay before fusing in the next few hundred seconds, so at the end of nucleosynthesis there are about seven protons to every neutron, and almost all the neutrons are in Helium-4 nuclei. The sequence of these reaction chains is shown on the image.

One feature of BBN is that the physical laws and constants that govern the behavior of matter at these energies are very well understood, and hence BBN lacks some of the speculative uncertainties that characterize earlier periods in the life of the universe. Another feature is that the process of nucleosynthesis is determined by conditions at the start of this phase of the life of the universe, and proceeds independently of what happened before.

As the universe expands, it cools. Free neutrons and protons are less stable than helium nuclei, and the protons and neutrons have a strong tendency to form helium-4. However, forming helium-4 requires the intermediate step of forming deuterium. Before nucleosynthesis began, the temperature was high enough for many photons to have energy greater than the binding energy of deuterium; therefore any deuterium that was formed was immediately destroyed (a situation known as the deuterium bottleneck). Hence, the formation of helium-4 is delayed until the universe became cool enough for deuterium to survive (at about T = 0.1 MeV); after which there was a sudden burst of element formation. However, very shortly thereafter, around twenty minutes after the Big Bang, the temperature and density became too low for any significant fusion to occur. At this point, the elemental abundances were nearly fixed, and the only changes were the result of the radioactive decay of the two major unstable products of BBN, tritium and beryllium-7.

History of Theory

The history of Big Bang nucleosynthesis began with the calculations of Ralph Alpher in the 1940s. Alpher published the Alpher–Bethe–Gamow paper that outlined the theory of light-element production in the early universe.

During the 1970s, there was a major puzzle in that the density of baryons as calculated by Big Bang nucleosynthesis was much less than the observed mass of the universe based on measurements of galaxy rotation curves and galaxy cluster dynamics. This puzzle was resolved in large part by postulating the existence of dark matter.

Heavy Elements

Big Bang nucleosynthesis produced no elements heavier than lithium, due to a bottleneck: the absence of a stable nucleus with 8 or 5 nucleons. This deficit of larger atoms also limited the amounts of lithium-7 produced during BBN. In stars, the bottleneck is passed by triple collisions of helium-4 nuclei, producing carbon (the triple-alpha process). However, this process is very slow and requires much higher densities, taking tens of thousands of years to convert a significant amount of helium to carbon in stars, and therefore it made a negligible contribution in the minutes following the Big Bang.

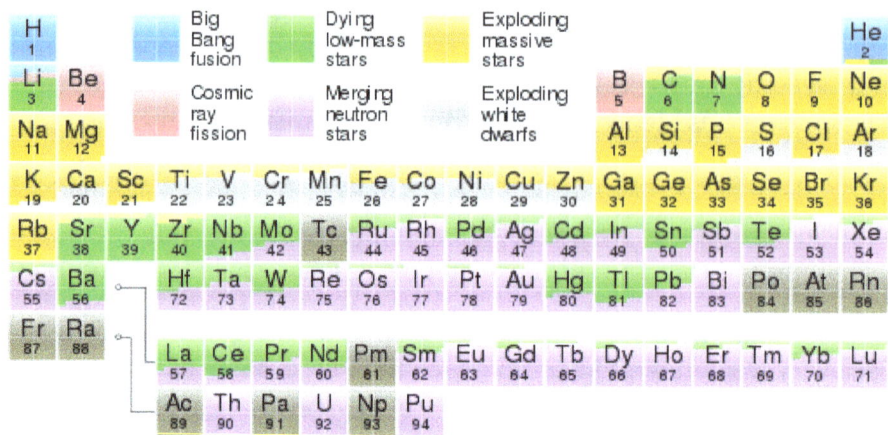

A version of the periodic table indicating the origins – including big bang nucleosynthesis – of the elements. All elements above 103 (lawrencium) are also manmade and are not included.

The predicted abundance of CNO isotopes produced in Big Bang nucleosynthesis is expected to be on the order of 10^{-15} that of H, making them essentially undetectable and negligible. Indeed, none of these primordial isotopes of the elements from lithium to oxygen have yet been detected, although those of beryllium and boron may be able to be detected in the future. So far, the only stable nuclides known experimentally to have been made before or during Big Bang nucleosynthesis are protium, deuterium, helium-3, helium-4, and lithium-7.

Helium-4

Big Bang nucleosynthesis predicts a primordial abundance of about 25% helium-4 by mass, irrespective of the initial conditions of the universe. As long as the universe was hot enough for protons and neutrons to transform into each other easily, their ratio, determined solely by their relative masses, was about 1 neutron to 7 protons (allowing for some decay of neutrons into protons). Once it was cool enough, the neutrons quickly bound with an equal number of protons to form first deuterium, then helium-4. Helium-4 is very stable and is nearly the end of this chain if it runs for only a short time, since helium neither decays nor combines easily to form heavier nuclei (since there are no stable nuclei with mass numbers of 5 or 8, helium does not combine easily with either protons, or with itself). Once temperatures are lowered, out of every 16 nucleons (2 neutrons and 14 protons), 4 of these (25% of the total particles and total mass) combine quickly into one helium-4 nucleus. This produces one helium for every 12 hydrogens, resulting in a universe that is a little over 8% helium by number of atoms, and 25% helium by mass.

One analogy is to think of helium-4 as ash, and the amount of ash that one forms when one completely burns a piece of wood is insensitive to how one burns it. The resort to the BBN theory of the helium-4 abundance is necessary as there is far more helium-4 in the universe than can be explained by stellar nucleosynthesis. In addition, it provides an important test for the Big Bang theory. If the observed helium abundance is significantly different from 25%, then this would pose a serious challenge to the theory. This would particularly be the case if the early helium-4 abundance was much smaller than 25% because it is hard to destroy helium-4. For a few years during the mid-1990s, observations suggested that this might be the case, causing astrophysicists to talk about a Big Bang nucleosynthetic crisis, but further observations were consistent with the Big Bang theory.

Deuterium

Deuterium is in some ways the opposite of helium-4 in that while helium-4 is very stable and very difficult to destroy, deuterium is only marginally stable and easy to destroy. The temperatures, time, and densities were sufficient to combine a substantial fraction of the deuterium nuclei to form helium-4 but insufficient to carry the process further using helium-4 in the next fusion step. BBN did not convert all of the deuterium in the universe to helium-4 due to the expansion that cooled the universe and reduced the density and so, cut that conversion short before it could proceed any further. One consequence of this is that unlike helium-4, the amount of deuterium is very sensitive to initial conditions. The denser the initial universe was, the more deuterium would be converted to helium-4 before time ran out, and the less deuterium would remain.

There are no known post-Big Bang processes which can produce significant amounts of deuterium. Hence observations about deuterium abundance suggest that the universe is not infinitely old, which is in accordance with the Big Bang theory.

During the 1970s, there were major efforts to find processes that could produce deuterium, but those revealed ways of producing isotopes other than deuterium. The problem was that while the concentration of deuterium in the universe is consistent with the Big Bang model as a whole, it is too high to be consistent with a model that presumes that most of the universe is composed of protons and neutrons. If one assumes that all of the universe consists of protons and neutrons, the density of the universe is such that much of the currently observed deuterium would have been burned into helium-4. The standard explanation now used for the abundance of deuterium is that the universe does not consist mostly of baryons, but that non-baryonic matter (also known as dark matter) makes up most of the mass of the universe. This explanation is also consistent with calculations that show that a universe made mostly of protons and neutrons would be far more *clumpy* than is observed.

It is very hard to come up with another process that would produce deuterium other than by nuclear fusion. Such a process would require that the temperature be hot enough to produce deuterium, but not hot enough to produce helium-4, and that this process should immediately cool to non-nuclear temperatures after no more than a few minutes. It would also be necessary for the deuterium to be swept away before it reoccurs.

Producing deuterium by fission is also difficult. The problem here again is that deuterium is very unlikely due to nuclear processes, and that collisions between atomic nuclei are likely to result either in the fusion of the nuclei, or in the release of free neutrons or alpha particles. During the 1970s, cosmic ray spallation was proposed as a source of deuterium. That theory failed to account for the abundance of deuterium, but led to explanations of the source of other light elements.

Measurements and Status of Theory

The theory of BBN gives a detailed mathematical description of the production of the light "elements" deuterium, helium-3, helium-4, and lithium-7. Specifically, the theory yields precise quantitative predictions for the mixture of these elements, that is, the primordial abundances at the end of the big-bang.

In order to test these predictions, it is necessary to reconstruct the primordial abundances as faithfully as possible, for instance by observing astronomical objects in which very little stellar nucleosynthesis has taken place (such as certain dwarf galaxies) or by observing objects that are very far away, and thus can be seen in a very early stage of their evolution (such as distant quasars).

As noted above, in the standard picture of BBN, all of the light element abundances depend on the amount of ordinary matter (baryons) relative to radiation (photons). Since the universe is presumed to be homogeneous, it has one unique value of the baryon-to-photon ratio. For a long time, this meant that to test BBN theory against observations one had to ask: can *all* of the light element observations be explained with a *single value* of the baryon-to-photon ratio? Or more precisely, allowing for the finite precision of both the predictions and the observations, one asks: is there some *range* of baryon-to-photon values which can account for all of the observations?

More recently, the question has changed: Precision observations of the cosmic microwave background radiation with the Wilkinson Microwave Anisotropy Probe (WMAP) and Planck give an independent value for the baryon-to-photon ratio. Using this value, are the BBN predictions for the abundances of light elements in agreement with the observations?

The present measurement of helium-4 indicates good agreement, and yet better agreement for helium-3. But for lithium-7, there is a significant discrepancy between BBN and WMAP/Planck, and the abundance derived from Population II stars. The discrepancy is a factor of 2.4−4.3 below the theoretically predicted value and is considered a problem for the original models, that have resulted in revised calculations of the standard BBN based on new nuclear data, and to various reevaluation proposals for primordial proton-proton nuclear reactions, especially the abundances of $^7Be + n \rightarrow {^7Li} + p$, versus $^7Be + {^2H} \rightarrow {^8Be} + p$.

Non-standard Scenarios

In addition to the standard BBN scenario there are numerous non-standard BBN scenarios. These should not be confused with non-standard cosmology: a non-standard BBN scenario assumes that the Big Bang occurred, but inserts additional physics in order to see how this affects elemental abundances. These pieces of additional physics include relaxing or removing the assumption of homogeneity, or inserting new particles such as massive neutrinos.

There have been, and continue to be, various reasons for researching non-standard BBN. The first, which is largely of historical interest, is to resolve inconsistencies between BBN predictions and observations. This has proved to be of limited usefulness in that the inconsistencies were resolved by better observations, and in most cases trying to change BBN resulted in abundances that were more inconsistent with observations rather than less. The second reason for researching non-standard BBN, and largely the focus of non-standard BBN in the early 21st century, is to use BBN to place limits on unknown or speculative physics. For example, standard BBN assumes that no exotic hypothetical particles were involved in BBN. One can insert a hypothetical particle (such as a massive neutrino) and see what has to happen before BBN predicts abundances that are very different from observations. This has been done to put limits on the mass of a stable tau neutrino.

Absolute Time and Space

Absolute space and time is a concept in physics and philosophy about the properties of the universe. In physics, absolute space and time may be a preferred frame.

Before Newton

A version of the concept of absolute space (in the sense of a preferred frame) can be seen in Aristotelian physics. Robert S. Westman writes that "whiff" of absolute space can be observed in Copernicus De revolutionibus orbium coelestium, where he exploits the concept of immobile sphere of stars.

Newton

Originally introduced by Sir Isaac Newton in *Philosophiæ Naturalis Principia Mathematica*, the concepts of absolute time and space provided a theoretical foundation that facilitated Newtonian mechanics. According to Newton, absolute time and space respectively are independent aspects of objective reality:

Absolute, true and mathematical time, of itself, and from its own nature flows equably without regard to anything external, and by another name is called duration: relative, apparent and common time, is some sensible and external (whether accurate or unequable) measure of duration by the means of motion, which is commonly used instead of true time ...

According to Newton, absolute time exists independently of any perceiver and progresses at a consistent pace throughout the universe. Unlike relative time, Newton believed absolute time was imperceptible and could only be understood mathematically. According to Newton, humans are only capable of perceiving relative time, which is a measurement of perceivable objects in motion (like the Moon or Sun). From these movements, we infer the passage of time.

Absolute space, in its own nature, without regard to anything external, remains always similar and immovable. Relative space is some movable dimension or measure of the absolute spaces; which our senses determine by its position to bodies: and which is vulgarly taken for immovable space ... Absolute motion is the translation of a body from one absolute place into another: and relative motion, the translation from one relative place into another ...

—Isaac Newton

These notions imply that absolute space and time do not depend upon physical events, but are a backdrop or stage setting within which physical phenomena occur. Thus, every object has an absolute state of motion relative to absolute space, so that an object must be either in a state of absolute rest, or moving at some absolute speed. To support his views, Newton provided some empirical examples: according to Newton, a solitary rotating sphere can be inferred to rotate about its axis relative to absolute space by observing the bulging of its equator, and a solitary pair of spheres tied by a rope can be inferred to be in absolute rotation about their center of gravity (barycenter) by observing the tension in the rope.

Absolute time and space continue to be used in classical mechanics, but modern formulations by authors such as Walter Noll and Clifford Truesdell go beyond the linear algebra of elastic moduli to use topology and functional analysis for non-linear field theories.

Differing Views

Historically, there have been differing views on the concept of absolute space and time. Gottfried Leibniz was of the opinion that space made no sense except as the relative location of bodies, and time made no sense except as the relative movement of bodies. George Berkeley suggested that, lacking any point of reference, a sphere in an otherwise empty universe could not be conceived to rotate, and a pair of spheres could be conceived to rotate relative to one another, but not to rotate about their center of gravity, an example later raised by Albert Einstein in his development of general relativity.

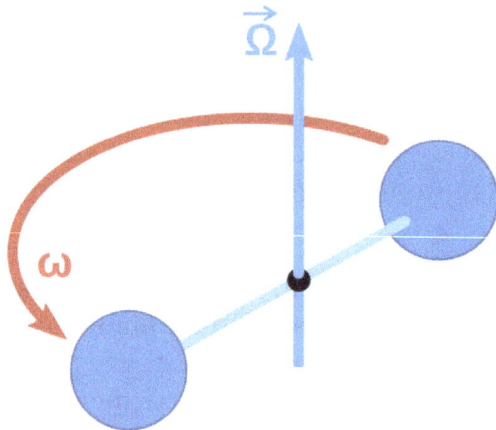

Two spheres orbiting around an axis. The spheres are distant enough for their effects on each other to be ignored, and they are held together by a rope. The rope is under tension if the bodies are rotating relative to absolute space according to Newton, or because they rotate relative to the universe itself according to Mach, or because they rotate relative to local geodesics according to general relativity.

A more recent form of these objections was made by Ernst Mach. Mach's principle proposes that mechanics is entirely about relative motion of bodies and, in particular, mass is an expression of such relative motion. So, for example, a single particle in a universe with no other bodies would have zero mass. According to Mach, Newton's examples simply illustrate relative rotation of spheres and the bulk of the universe.

When, accordingly, we say that a body preserves unchanged its direction and velocity *in space*, our assertion is nothing more or less than an abbreviated reference to *the entire universe*.

—Ernst Mach; as quoted by Ciufolini and Wheeler: *Gravitation and Inertia*, p. 387

These views opposing absolute space and time may be seen from a modern stance as an attempt to introduce operational definitions for space and time, a perspective made explicit in the special theory of relativity.

Even within the context of Newtonian mechanics, the modern view is that absolute space is unnecessary. Instead, the notion of inertial frame of reference has taken precedence, that is, a preferred *set* of frames of reference that move uniformly with respect to one another. The laws of physics transform from one inertial frame to another according to Galilean relativity, leading to the following objections to absolute space, as outlined by Milutin Blagojević:

- The existence of absolute space contradicts the internal logic of classical mechanics since, according to Galilean principle of relativity, none of the inertial frames can be singled out.

- Absolute space does not explain inertial forces since they are related to acceleration with respect to any one of the inertial frames.

- Absolute space acts on physical objects by inducing their resistance to acceleration but it cannot be acted upon.

Newton himself recognized the role of inertial frames.

The motions of bodies included in a given space are the same among themselves, whether that space is at rest or moves uniformly forward in a straight line.

As a practical matter, inertial frames often are taken as frames moving uniformly with respect to the fixed stars.

Special Relativity

The concepts of space and time were separate in physical theory prior to the advent of special relativity theory, which connected the two and showed both to be dependent upon the reference frame's motion. In Einstein's theories, the ideas of absolute time and space were superseded by the notion of spacetime in special relativity, and curved spacetime in general relativity.

Absolute simultaneity refers to the concurrence of events in time at different locations in space in a manner agreed upon in all frames of reference. The theory of relativity does not have a concept of absolute time because there is a relativity of simultaneity. An event that is simultaneous with another event in one frame of reference may be in the past or future of that event in a different frame of reference, which negates absolute simultaneity.

Einstein

Quoted below from his later papers, Einstein identified the term aether with "properties of space", a terminology that is not widely used. Einstein stated that in general relativity the "aether" is not absolute anymore, as the geodesic and therefore the structure of spacetime depends on the presence of matter.

To deny the ether is ultimately to assume that empty space has no physical qualities whatever. The fundamental facts of mechanics do not harmonize with this view. For the mechanical behaviour of a corporeal system hovering freely in empty space depends not only on relative positions (distances) and relative velocities, but also on its state of rotation, which physically may be taken as a characteristic not appertaining to the system in itself. In order to be able to look upon the rotation of the system, at least formally, as something real, Newton objectivises space. Since he classes his absolute space together with real things, for him rotation relative to an absolute space is also something real. Newton might no less well have called his absolute space "Ether"; what is essential is merely that besides observable objects, another thing, which is not perceptible, must be looked upon as real, to enable acceleration or rotation to be looked upon as something real.

—Albert Einstein, Ether and the Theory of Relativity (1920)

Because it was no longer possible to speak, in any absolute sense, of simultaneous states at different locations in the aether, the aether became, as it were, four-dimensional, since there was no objective way of ordering its states by time alone. According to special relativity too, the aether was absolute, since its influence on inertia and the propagation of light was thought of as being itself independent of physical influence....The theory of relativity resolved this problem by establishing the behaviour of the electrically neutral point-mass by the law of the geodetic line, according to which inertial and gravitational effects are no longer considered as separate. In doing so, it attached characteristics to the aether which vary from point to point, determining the metric and the dynamic behaviour of material points, and determined, in their turn, by physical factors, namely the distribution of mass/energy. Thus the aether of general relativity differs from those of classical mechanics and special relativity in that it is not 'absolute' but determined, in its locally variable characteristics, by ponderable matter.

—*Albert Einstein, Über den Äther (1924)*

General Relativity

Special relativity eliminates absolute time (although Gödel and others suspect absolute time may be valid for some forms of general relativity) and general relativity further reduces the physical scope of absolute space and time through the concept of geodesics. There appears to be absolute space in relation to the distant stars because the local geodesics eventually channel information from these stars, but it is not necessary to invoke absolute space with respect to any system's physics.

De Revolutionibus Orbium Coelestium

De revolutionibus orbium coelestium (*On the Revolutions of the Heavenly Spheres*) is the seminal work on the heliocentric theory of the Renaissance astronomer Nicolaus Copernicus (1473–1543). The book, first printed in 1543 in Nuremberg, Holy Roman Empire, offered an alternative model of the universe to Ptolemy's geocentric system, which had been widely accepted since ancient times.

History

Heliocentric model of the solar system in Copernicus' manuscript

Copernicus initially outlined his system in a short, untitled, anonymous manuscript that he distributed to several friends, referred to as the *Commentariolus*. A physician's library list dating to 1514 includes a manuscript whose description matches the *Commentariolus*, so Copernicus must have begun work on his new system by that time. Most historians believe that he wrote the *Commentariolus* after his return from Italy, possibly only after 1510. At this time, Copernicus anticipated that he could reconcile the motion of the Earth with the perceived motions of the planets easily, with fewer motions than were necessary in the *Alfonsine Tables*, the version of the Ptolemaic system current at the time. In particular, the heliocentric Copernican model made use of the Urdi Lemma developed in the 13th century by Mu'ayyad al-Din al-'Urdi, the first of the Maragha astronomers to develop a non-Ptolemaic model of planetary motion.

Observations of Mercury by Bernhard Walther (1430–1504) of Nuremberg, a pupil of Regiomontanus, were made available to Copernicus by Johannes Schöner, 45 observations in total, 14 of them with longitude and latitude. Copernicus used three of them in *De revolutionibus*, giving only longitudes, and erroneously attributing them to Schöner. Copernicus' values differed slightly from the ones published by Schöner in 1544 in *Observationes XXX annorum a I. Regiomontano et B. Walthero Norimbergae habitae, [4°, Norimb. 1544]*.

A manuscript of *De revolutionibus* in Copernicus' own hand has survived. After his death, it was given to his pupil, Rheticus, who for publication had only been given a copy without annotations. Via Heidelberg, it ended up in Prague, where it was rediscovered and studied in the 19th century. Close examination of the manuscript, including the different types of paper used, helped scholars construct an approximate timetable for its composition. Apparently Copernicus began by making a few astronomical observations to provide new data to perfect his models. He may have begun writing the book while still engaged in observations. By the 1530s a substantial part of the book was complete, but Copernicus hesitated to publish.

In 1539 Georg Joachim Rheticus, a young mathematician from Wittenberg, arrived in Frauenburg (Frombork) to study with him. Rheticus read Copernicus' manuscript and immediately wrote a non-technical summary of its main theories in the form of an open letter addressed to Schöner, his astrology teacher in Nürnberg; he published this letter as the *Narratio Prima* in Danzig in 1540. Rheticus' friend and mentor Achilles Gasser published a second edition of the *Narratio* in Basel in 1541. Due to its friendly reception, Copernicus finally agreed to publication of more of his main work—in 1542, a treatise on trigonometry, which was taken from the second book of the still unpublished *De revolutionibus*. Rheticus published it in Copernicus' name.

Under strong pressure from Rheticus, and having seen that the first general reception of his work had not been unfavorable, Copernicus finally agreed to give the book to his close friend, Bishop Tiedemann Giese, to be delivered to Rheticus in Wittenberg for printing by Johannes Petreius at Nürnberg (Nuremberg). It was published just before Copernicus' death, in 1543.

The book is dedicated to Pope Paul III in a preface that argues that mathematics, not physics, should be the basis for understanding and accepting his new theory.

De revolutionibus is divided into six "books" (sections or parts), following closely the layout of Ptolemy's *Almagest* which it updated and replaced:

- Book I chapters 1–11 are a general vision of the heliocentric theory, and a summarized exposition of his cosmology. The world (heavens) is spherical, as is the earth, and the land and water make a single globe. The celestial bodies, including the earth, have regular circular and everlasting movements. The earth rotates on its axis and around the sun. Answers to why the ancients thought the earth was central. The order of the planets around the sun and their periodicity. Chapters 12-14 give theorems for chord geometry as well as a table of chords.

- Book II describes the principles of spherical astronomy as a basis for the arguments developed in the following books and gives a comprehensive catalogue of the fixed stars.

- Book III describes his work on the precession of the equinoxes and treats the apparent movements of the Sun and related phenomena.

- Book IV is a similar description of the Moon and its orbital movements.

- Book V explains how to calculate the positions of the wandering stars based on the heliocentric model and gives tables for the five planets.

- Book VI deals with the digression in latitude from the ecliptic of the five planets.

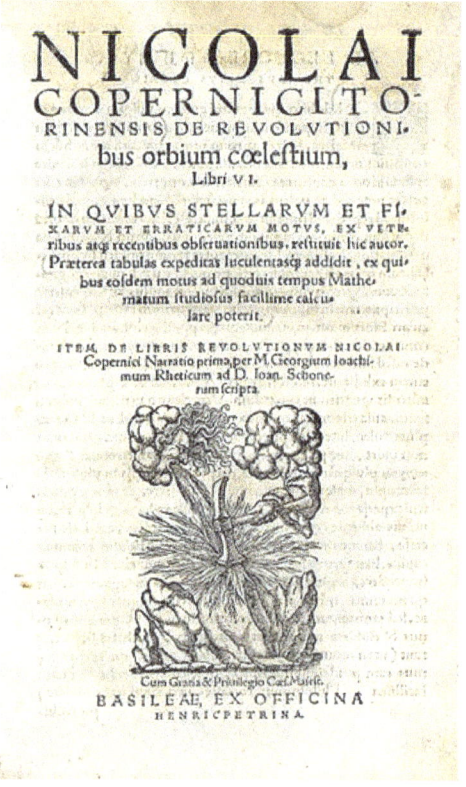

Title page, 2nd edition, Basel, *Officina Henricpetrina*, 1566

Copernicus argued that the universe comprised eight spheres. The outermost consisted of motionless, fixed stars, with the Sun motionless at the center. The known planets revolved about the Sun, each in its own sphere, in the order: Mercury, Venus, Earth, Mars, Jupiter, Saturn. The Moon, however, revolved in its sphere around the Earth. What appeared to be the daily revolution of the Sun and fixed stars around the Earth was actually the Earth's daily rotation on its own axis.

Copernicus adhered to one of the standard beliefs of his time, namely that the motions of celestial bodies must be composed of uniform circular motions. For this reason, he was unable to account for the observed apparent motion of the planets without retaining a complex system of epicycles similar to those of the Ptolemaic system. Despite Copernicus' adherence to this aspect of ancient astronomy, his radical shift from a geocentric to a heliocentric cosmology was a serious blow to Aristotle's science—and helped usher in the Scientific Revolution.

Ad Lectorem

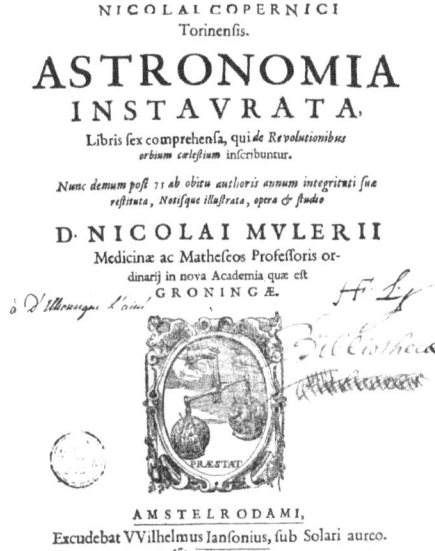

NICOLAI COPERNICI
Torinenfis.

ASTRONOMIA
INSTAVRATA,

Libris fex comprehenfa, qui *de Revolutionibus orbium cæleftium* infcribuntur.

Nunc demum poft 75 ab obitu authoris annum integritati fuæ reftituta, Notifque illuftrata, opera & ftudio

D· NICOLAI MVLERII

Medicinæ ac Mathefeos Profefforis or-
dinarij in nova Academia quæ eft
GRONINGÆ.

PRÆSTAT

AMSTELRODAMI,

Excudebat VVilhelmus Ianfonius, fub Solari aureo.
Ano M. D C. XVII.

Title page, 3rd ed., Amsterdam, Nicolaus Mulerius, publisher, 1617

Rheticus left Nürnberg to take up his post as professor in Leipzig. The Lutheran preacher Andreas Osiander had taken over the task of supervising the printing and publication. In an effort to reduce the controversial impact of the book Osiander added his own unsigned letter *Ad lectorem de hypothesibus huius operis* (*To the reader concerning the hypotheses of this work*) printed in front of Copernicus' preface which was a dedicatory letter to Pope Paul III and which kept the title "Praefatio authoris" (to acknowledge that the unsigned letter was not by the book's author).

Osiander's letter stated that Copernicus' system was mathematics intended to aid computation and not an attempt to declare literal truth:

it is the duty of an astronomer to compose the history of the celestial motions through careful and expert study. Then he must conceive and devise the causes of these motions or hypotheses about them. Since he cannot in any way attain to the true causes, he will adopt whatever suppositions enable the motions to be computed correctly ... The present author has performed both these duties excellently. For these hypotheses need not be true nor even probable. On the contrary, if they provide a calculus consistent with the observations, that alone is enough ... For this art, it is quite clear, is completely and absolutely ignorant of the causes of the apparent [movement of the heavens]. And if any causes are devised by the imagination, as indeed very many are, they are not put forward to convince anyone that they are true, but merely to provide a reliable basis for computation. However, since different hypotheses are sometimes offered for one and the same ... the astronomer will take as his first choice that hypothesis which is the easiest to grasp. The philosopher will perhaps rather seek the semblance of the truth. But neither of them will understand or state anything certain, unless it has been divinely revealed to him ... Let no one expect anything certain from astronomy, which cannot furnish it, lest he accept as the truth ideas conceived for another purpose, and depart this study a greater fool than when he entered.

As even Osiander's defenders point out, the *Ad lectorem* "expresses views on the aim and nature of scientific theories at variance with Copernicus' claims for his own theory".

Many view Osiander's letter as a betrayal of science and Copernicus, and an attempt to pass his own thoughts off as those of the book's author. An example of this type of claim can be seen in the *Catholic Encyclopedia*, which states "Fortunately for him [the dying Copernicus], he could not see what Osiander had done. This reformer, knowing the attitude of Luther and Melanchthon against the heliocentric system ... without adding his own name, replaced the preface of Copernicus by another strongly contrasting in spirit with that of Copernicus."

While Osiander's motives behind the letter have been questioned by many, he has been defended by historian Bruce Wrightsman, who points out he was not an enemy of science. Osiander had many scientific connections including "Johannes Schoner, Rheticus's teacher, whom Osiander recommended for his post at the Nurnberg Gymnasium; Peter Apian of Ingolstadt University; Hieronymous Schreiber...Joachim Camerarius...Erasmus Reinhold...Joachim Rheticus...and finally, Hieronymous Cardan."

The historian Wrightsman put forward that Osiander did not sign the letter because he "was such a notorious [Protestant] reformer whose name was well-known and infamous among Catholics", so that signing would have likely caused negative scrutiny of the work of Copernicus (a loyal Catholic canon and scholar). Copernicus himself had communicated to Osiander his "own fears that his work would be scrutinized and criticized by the 'peripatetics and theologians'," and he had already been in trouble with his bishop, Johannes Dantiscus, on account of his former relationship with his mistress and friendship with Dantiscus's enemy and suspected heretic, Alexander Scultetus. It was also possible that Protestant Nurnberg could fall to the forces of the Holy Roman Emperor and since "the books of hostile theologians could be burned...why not scientific works with the names of hated theologians affixed to them?" Wrightsman also holds that this is why Copernicus did not mention his top student, Rheticus (a Lutheran) in the book's dedication to the Pope.

Osiander's interest in astronomy was theological, hoping for "improving the chronology of historical events and thus providing more accurate apocalyptic interpretations of the Bible... [he shared in] the general awareness that the calendar was not in agreement with astronomical movement and therefore, needed to be corrected by devising better models on which to base calculations." In an era before the telescope, Osiander (like most of the era's mathematical astronomers) attempted to bridge the "fundamental incompatibility between Ptolemaic astronomy and Aristotlian physics, and the need to preserve both", by taking an 'instrumentalist' position. Only the handful of "Philosophical purists like the Averroists... demanded physical consistency and thus sought for realist models."

Copernicus was hampered by his insistence on preserving the idea that celestial bodies had to travel in perfect spheres – he "was still attached to classical ideas of circular motion around deferents and epicycles, and spheres." This was particularly troubling concerning the Earth because he "attached the Earth's axis rigidly to a Sun-centered sphere. The unfortunate consequence was that the terrestrial rotation axis then maintained the same inclination with respect to the Sun as the sphere turned, eliminating the seasons." To explain the seasons, he had to propose a third motion, "an annual contrary conical sweep of the terrestrial axis". It was not until the Great Comet of 1577, which moved as if there were no spheres to crash through, did the idea come under question. In 1609, Kepler fixed Copernicus' theory by stating that the planets orbit the sun not in circles, but ellipses. Only after Kepler's refinement of Copernicus' theory was the need for deferents and epicycles abolished.

In his work, Copernicus "used conventional, hypothetical devices like epicycles...as all astronomers had done since antiquity. ...hypothetical constructs solely designed to 'save the phenomena' and aid computation". Ptolemy's theory contained a hypothesis about the epicycle of Venus that was viewed as absurd if seen as anything other than a geometrical device (its brightness and distance should have varied greatly, but they don't). "In spite of this defect in Ptolemy's theory, Copernicus' hypothesis predicts approximately the same variations." Because of the use of similar terms and similar deficiencies, Osiander could see "little technical or physical truth-gain" between one system and the other. It was this attitude towards technical astronomy that had allowed it to "function since antiquity, despite its inconsistencies with the principles of physics and the philosophical objections of Averroists."

Writing *Ad lectorem*, Osiander was influenced by Pico della Mirandola's idea that humanity "orders [an intellectual] cosmos out of the chaos of opinions." From Pico's writings, Osiander "learned to extract and synthesize insights from many sources without becoming the slavish follower of any of them." The effect of Pico on Osiander was tempered by the influence of Nicholas of Cusa's and his idea of *coincidentia oppositorum*. Rather than having Pico's focus on human effort, Osiander followed Cusa's idea that understanding the Universe and its Creator only came from divine inspiration rather than intellectual organization. From these influences, Osiander held that in the area of philosophical speculation and scientific hypothesis there are "no heretics of the intellect", but when one gets past speculation into truth-claims the Bible is the ultimate measure. By holding Copernicianism was mathematical speculation, Osiander held that it would be silly to hold it up against the accounts of the Bible.

Pico's influence on Osiander did not escape Rheticus, who reacted strongly against the *Ad lectorem*. As historian Robert S. Westman puts it, "The more profound source of Rheticus's ire however, was Osiander's view of astronomy as a disciple fundamentally incapable of knowing anything with certainty. For Rheticus, this extreme position surely must have resonated uncomfortably with Pico della Mirandola's attack on the foundations of divinatory astrology."

In his *Disputations*, Pico had made a devastating attack on astrology. Because those who were making astrological predictions relied on astronomers to tell them where the planets were, they also became a target. Pico held that since astronomers who calculate planetary positions could not agree among themselves, how were they to be held as reliable? While Pico could bring into concordance writers like Aristotle, Plato, Plotinus, Averroes, Avicenna, and Aquinas, the lack of consensus he saw in astronomy was a proof to him of its fallibility alongside astrology. Pico pointed out that the astronomers' instruments were imprecise and any imperfection of even a degree made them worthless for astrology, people should not trust astrologists because they should not trust the numbers from astronomers. Pico pointed out that astronomers couldn't even tell where the sun appeared in the order of the planets as they orbited the earth (some put it close to the moon, others among the planets). How, Pico asked, could astrologists possibly claim they could read what was going on when the astronomers they relied on could offer no precision on even basic questions?

As Westman points out, to Rheticus "it would seem that Osiander now offered new grounds for endorsing Pico's conclusions: not merely was the disagreement among astronomers grounds for mistrusting the sort of knowledge that they produced, but now Osiander proclaimed that astronomers might construct a world deduced from (possibly) false premises. Thus the conflict between Piconian skepticism and secure principles for the science of the stars was built right into the complex dedicatory apparatus of *De Revolutionibus* itself." According to the notes of Michael Maestlin, "Rheticus...became embroiled in a very bitter wrangle with the printer [over the Ad lectorem]. Rheticus...suspected Osiander had prefaced the work; if he knew this for certain, he declared, he would rough up the fellow so violently that in future he would mind his own business."

Objecting to the *Ad lectorem*, Tiedemann Giese urged the Nuremberg city council to issue a correction, but this was not done, and the matter was forgotten. Jan Broscius, a supporter of Copernicus, also despaired of the *Ad lectorem*, writing "Ptolemy's hypothesis is the earth rests. Copernicus' hypothesis is that the earth is in motion. Can either, therefore, be true? ... Indeed, Osiander deceives much with that preface of his ... Hence, someone may well ask: How is one to know which hypothesis is truer, the Ptolemaic or the Copernican?"

Petreius had sent a copy to Hieronymus Schreiber, an astronomer from Nürnberg who had substituted for Rheticus as professor of mathematics in Wittenberg while Rheticus was in Nürnberg supervising the printing. Schreiber, who died in 1547, left in his copy of the book a note about Osiander's authorship. Via Michael Mästlin, this copy came to Johannes Kepler, who discovered what Osiander had done and methodically demonstrated that Osiander had indeed added the foreword. The most knowledgeable astronomers of the time had realized that the foreword was Osiander's doing.

Owen Gingerich gives a slightly different version: Kepler knew of Osiander's authorship since he had read about it in one of Schreiber's annotations in his copy of *De Revolutionibus*; Maestlin learned of the fact from Kepler. Indeed, Maestlin perused Kepler's book, up to the point of leaving a few annotations in it. However, Maestlin already suspected Osiander, because he had bought his *De revolutionibus* from the widow of Philipp Apian; examining his books, he had found a note attributing the introduction to Osiander.

Johannes Praetorius (1537–1616), who learned of Osiander's authorship from Rheticus during a visit to him in Kraków, wrote Osiander's name in the margin of the foreword in his copy of *De revolutionibus*.

All three early editions of *De revolutionibus* included Osiander's foreword.

Reception

Even before the 1543 publication of *De revolutionibus*, rumors circulated about its central theses. Martin Luther is quoted as saying in 1539:

People gave ear to an upstart astrologer who strove to show that the earth revolves, not the heavens or the firmament, the sun and the moon ... This fool wishes to reverse the entire science of astronomy; but sacred Scripture tells us [Joshua 10:13] that Joshua commanded the sun to stand still, and not the earth.

When the book was finally published, demand was low, with an initial print run of 400 failing to sell out. Copernicus had made the book extremely technical, unreadable to all but the most advanced astronomers of the day, allowing it to disseminate into their ranks before stirring great controversy. And, like Osiander, contemporary mathematicians and astronomers encouraged its audience to view it as a useful mathematical fiction with no physical reality, thereby somewhat shielding it from accusations of blasphemy.

Among some astronomers, the book "at once took its place as a worthy successor to the *Almagest* of Ptolemy, which had hitherto been the Alpha and Omega of astronomers". Erasmus Reinhold hailed the work in 1542 and by 1551 had developed the *Prutenic Tables* ("Prussian Tables"; Latin: *Tabulae prutenicae*; German: *Preußische Tafeln*) using Copernicus' methods. The *Prutenic Tables*, published in 1551, were used as a basis for the calendar reform instituted in 1582 by Pope Gregory XIII. They were also used by sailors and maritime explorers, whose 15th-century predecessors had used Regiomontanus' *Table of the Stars*. In England, Robert Recorde, John Dee, Thomas Digges and William Gilbert were among those who adopted his position; in Germany, Christian Wurstisen, Christoph Rothmann and Michael Mästlin, the teacher of Johannes Kepler; in Italy, Giambattista Benedetti and Giordano Bruno whilst Franciscus Patricius accepted the rotation of the earth. In Spain, rules published in 1561 for the curriculum of the University of Salamanca gave students the choice between studying Ptolemy or Copernicus. One of those students, Diego de Zúñiga, published an acceptance of Copernican theory in 1584.

Very soon, nevertheless, Copernicus' theory was attacked with Scripture and with the common Aristotelian proofs. In 1549 Melanchthon, Luther's principal lieutenant, wrote against Copernicus, pointing to the theory's apparent conflict with Scripture and advocating that "severe measures" be taken to restrain the impiety of Copernicans. The works of Copernicus and Zúñiga—the latter for asserting that *De revolutionibus* was compatible with Catholic faith—were placed on the Index of Forbidden Books by a decree of the Sacred Congregation of March 5, 1616 (more than 70 years after Copernicus' publication):

This Holy Congregation has also learned about the spreading and acceptance by many of the false Pythagorean doctrine, altogether contrary to the Holy Scripture, that the earth moves and the sun is motionless, which is also taught by Nicholaus Copernicus' *De revolutionibus orbium coelestium* and by Diego de Zúñiga's *In Job* ... Therefore, in order that this opinion may not creep any further to the prejudice of Catholic truth, the Congregation has decided that the books by Nicolaus Copernicus [*De revolutionibus*] and Diego de Zúñiga [*In Job*] be suspended until corrected.

De revolutionibus was not formally banned but merely withdrawn from circulation, pending "corrections" that would clarify the theory's status as hypothesis. Nine sentences that represented the heliocentric system as certain were to be omitted or changed. After these corrections were prepared and formally approved in 1620 the reading of the book was permitted. But the book was never reprinted with the changes and was available in Catholic jurisdictions only to suitably qualified scholars, by special request. It remained on the Index until 1758, when Pope Benedict XIV (1740–58) removed the uncorrected book from his revised Index.

Census of Copies

Arthur Koestler described *De revolutionibus* as "*The Book That Nobody Read*" saying the book "was and is an all-time worst seller", despite the fact that it was reprinted four times. Owen Gingerich, a writer on both Nicolaus Copernicus and Johannes Kepler, disproved this after a 35-year project to examine every surviving copy of the first two editions. Gingerich showed that nearly all the leading mathematicians and astronomers of the time owned and read the book; however, his analysis of the marginalia shows that they almost all ignored the cosmology at the beginning of the book and were only interested in Copernicus' new equant-free models of planetary motion in the later chapters. Also, Nicolaus Reimers in 1587 translated the book into German.

Gingerich's efforts and conclusions are recounted in *The Book Nobody Read*, published in 2004 by Walker & Co. His census included 276 copies of the first edition (by comparison, there are 228 extant copies of the First Folio of Shakespeare) and 325 copies of the second. The research behind this book earned its author the Polish government's Order of Merit in 1981. Due largely to Gingerich's scholarship, *De revolutionibus* has been researched and catalogued better than any other first-edition historic text except for the original Gutenberg Bible. One of the copies now resides at the Archives of the University of Santo Tomas in the Miguel de Benavides Library. In January 2017, a second-edition copy was stolen as part of a heist of rare books from Heathrow Airport and remains unrecovered.

Editions

- 1543, Nuremberg, by Johannes Petreius

- 1566, Basel, by Henricus Petrus

- 1617, Amsterdam, by Nicolaus Mulerius

- 1854, Warsaw, with Polish translation and the authentic preface by Copernicus.

- 1873, Thorn, German translation sponsored by the local *Coppernicus* Society, with all Copernicus' textual corrections given as footnotes.

Philosophiæ Naturalis Principia Mathematica

Philosophiæ Naturalis Principia Mathematica, often referred to as simply the *Principia*, is a work in three books by Isaac Newton, in Latin, first published 5 July 1687. After annotating and correcting his personal copy of the first edition, Newton published two further editions, in 1713 and 1726. The *Principia* states Newton's laws of motion, forming the foundation of classical mechanics; Newton's law of universal gravitation; and a derivation of Kepler's laws of planetary motion (which Kepler first obtained empirically). The *Principia* is considered as one of the most important works in the history of science.

The French mathematical physicist Alexis Clairaut assessed it in 1747: "The famous book of *Mathematical Principles of Natural Philosophy* marked the epoch of a great revolution in physics. The method followed by its illustrious author Sir Newton ... spread the light of mathematics on a science which up to then had remained in the darkness of conjectures and hypotheses." A more recent assessment has been that while acceptance of Newton's theories was not immediate, by the end of a century after publication in 1687, "no one could deny that" (out of the *Principia*) "a science had emerged that, at least in certain respects, so far exceeded anything that had ever gone before that it stood alone as the ultimate exemplar of science generally."

In formulating his physical theories, Newton developed and used mathematical methods now included in the field of calculus. But the language of calculus as we know it was largely absent from the *Principia*; Newton gave many of his proofs in a geometric form of infinitesimal calculus, based on limits of ratios of vanishing small geometric quantities. In a revised conclusion to the *Principia*, Newton used his expression that became famous, *Hypotheses non fingo* ("I formulate no hypotheses").

Expressed Aim and Topics Covered

Sir Isaac Newton (1643–1727) author of the *Principia*

In the preface of the *Principia*, Newton wrote:

[...] Rational Mechanics will be the science of motions resulting from any forces whatsoever, and of the forces required to produce any motions, accurately proposed and demonstrated [...] And therefore we offer this work as mathematical principles of philosophy. For all the difficulty of philosophy seems to consist in this—from the phenomena of motions to investigate the forces of Nature, and then from these forces to demonstrate the other phenomena [...]

The *Principia* deals primarily with massive bodies in motion, initially under a variety of conditions and hypothetical laws of force in both non-resisting and resisting media, thus offering criteria to decide, by observations, which laws of force are operating in phenomena that may be observed. It attempts to cover hypothetical or possible motions both of celestial bodies and of terrestrial projectiles. It explores difficult problems of motions perturbed by multiple attractive forces. Its third and final book deals with the interpretation of observations about the movements of planets and their satellites. It shows how astronomical observations prove the inverse square law of gravitation (to an accuracy that was high by the standards of Newton's time); offers estimates of relative masses for the known giant planets and for the Earth and the Sun; defines the very slow motion of the Sun relative to the solar-system barycenter; shows how the theory of gravity can account for irregularities in the motion of the Moon; identifies the oblateness of the figure of the Earth; accounts approximately for marine tides including phenomena of spring and neap tides by the perturbing (and varying) gravitational attractions of the Sun and Moon on the Earth's waters; explains the precession of the equinoxes as an effect of the gravitational attraction of the Moon on the Earth's equatorial bulge; and gives theoretical basis for numerous phenomena about comets and their elongated, near-parabolic orbits.

The opening sections of the *Principia* contain, in revised and extended form, nearly all of the content of Newton's 1684 tract *De motu corporum in gyrum*.

The *Principia* begin with "Definitions" and "Axioms or Laws of Motion", and continues in three books:

Book 1, *De Motu Corporum*

Book 1, subtitled *De motu corporum* (*On the motion of bodies*) concerns motion in the absence of any resisting medium. It opens with a mathematical exposition of "the method of first and last ratios", a geometrical form of infinitesimal calculus.

The second section establishes relationships between centripetal forces and the law of areas now known as Kepler's second law (Propositions 1–3), and relates circular velocity and radius of path-curvature to radial force (Proposition 4), and relationships between centripetal forces varying as the inverse-square of the distance to the center and orbits of conic-section form (Propositions 5–10).

Propositions 11–31 establish properties of motion in paths of eccentric conic-section form including ellipses, and their relation with inverse-square central forces directed to a focus, and include Newton's theorem about ovals (lemma 28).

Propositions 43–45 are demonstration that in an eccentric orbit under centripetal force where the apse may move, a steady non-moving orientation of the line of apses is an indicator of an inverse-square law of force.

Book 1 contains some proofs with little connection to real-world dynamics. But there are also sections with far-reaching application to the solar system and universe:

Propositions 57–69 deal with the "motion of bodies drawn to one another by centripetal forces". This section is of primary interest for its application to the solar system, and includes Proposition 66 along with its 22 corollaries: here Newton took the first steps in the definition and study of the problem of the movements of three massive bodies subject to their mutually perturbing gravitational attractions, a problem which later gained name and fame (among other reasons, for its great difficulty) as the three-body problem.

Propositions 70–84 deal with the attractive forces of spherical bodies. The section contains Newton's proof that a massive spherically symmetrical body attracts other bodies outside itself as if all its mass were concentrated at its centre. This fundamental result, called the Shell theorem, enables the inverse square law of gravitation to be applied to the real solar system to a very close degree of approximation.

Book 2

Part of the contents originally planned for the first book was divided out into a second book, which largely concerns motion through resisting mediums. Just as Newton examined consequences of different conceivable laws of attraction in Book 1, here he examines different conceivable laws of resistance; thus Section 1 discusses resistance in direct proportion to velocity, and Section 2 goes on to examine the implications of resistance in proportion to the square of velocity. Book 2 also discusses (in Section 5) hydrostatics and the properties of compressible fluids. The effects of air resistance on pendulums are studied in Section 6, along with Newton's account of experiments that he carried out, to try to find out some characteristics of air resistance in reality by observing the motions of pendulums under different conditions. Newton compares the resistance offered by a medium against motions of globes with different properties (material, weight, size). In Section 8, he derives rules to determine the speed of waves in fluids and relates them to the density and condensation(Proposition 48; this would become very important in acoustics). He assumes that these rules apply equally to light and sound and estimates that the speed of sound is around 1088 feet per second and can increase depending on the amount of water in air.

Less of Book 2 has stood the test of time than of Books 1 and 3, and it has been said that Book 2 was largely written on purpose to refute a theory of Descartes which had some wide acceptance before Newton's work (and for some time after). According to this Cartesian theory of vortices, planetary motions were produced by the whirling of fluid vortices that filled interplanetary space and carried the planets along with them. Newton wrote at the end of Book 2 his conclusion that the hypothesis of vortices was completely at odds with the astronomical phenomena, and served not so much to explain as to confuse them.

Book 3, *De Mundi Systemate*

Book 3, subtitled *De mundi systemate* (*On the system of the world*), is an exposition of many consequences of universal gravitation, especially its consequences for astronomy. It builds upon the propositions of the previous books, and applies them with further specificity than in Book 1 to the motions observed in the solar system. Here (introduced by Proposition 22, and continuing in Propositions 25–35) are developed several of the features and irregularities of the orbital motion of the Moon, especially the variation. Newton lists the astronomical observations on which he relies, and establishes in a stepwise manner that the inverse square law of mutual gravitation applies to solar system bodies, starting with the satellites of Jupiter and going on by stages to show that the law is of universal application. He also gives starting at Lemma 4 and Proposition 40 the theory of the motions of comets, for which much data came from John Flamsteed and Edmond Halley, and accounts for the tides, attempting quantitative estimates of the contributions of the Sun and Moon to the tidal motions; and offers the first theory of the precession of the equinoxes. Book 3 also considers the harmonic oscillator in three dimensions, and motion in arbitrary force laws.

In Book 3 Newton also made clear his heliocentric view of the solar system, modified in a somewhat modern way, since already in the mid-1680s he recognised the "deviation of the Sun" from the centre of gravity of the solar system. For Newton, "the common centre of gravity of the Earth, the Sun and all the Planets is to be esteem'd the Centre of the World", and that this centre "either is at rest, or moves uniformly forward in a right line". Newton rejected the second alternative after adopting the position that "the centre of the system of the world is immoveable", which "is acknowledg'd by all, while some contend that the Earth, others, that the Sun is fix'd in that centre". Newton estimated the mass ratios Sun:Jupiter and Sun:Saturn, and pointed out that these put the centre of the Sun usually a little way off the common center of gravity, but only a little, the distance at most "would scarcely amount to one diameter of the Sun".

Commentary on the *Principia*

The sequence of definitions used in setting up dynamics in the *Principia* is recognisable in many textbooks today. Newton first set out the definition of mass[6]

The quantity of matter is that which arises conjointly from its density and magnitude. A body twice as dense in double the space is quadruple in quantity. This quantity I designate by the name of body or of mass.

This was then used to define the "quantity of motion" (today called momentum), and the principle of inertia in which mass replaces the previous Cartesian notion of *intrinsic force*. This then set the stage for the introduction of forces through the change in momentum of a body. Curiously, for today's readers, the exposition looks dimensionally incorrect, since Newton does not introduce the dimension of time in rates of changes of quantities.

He defined space and time "not as they are well known to all". Instead, he defined "true" time and space as "absolute" and explained:

Only I must observe, that the vulgar conceive those quantities under no other notions but from the relation they bear to perceptible objects. And it will be convenient to distinguish them into absolute and relative, true and apparent, mathematical and common. [...] instead of absolute places and motions, we use relative ones; and that without any inconvenience in common affairs; but in philosophical discussions, we ought to step back from our senses, and consider things themselves, distinct from what are only perceptible measures of them.

To some modern readers it can appear that some dynamical quantities recognised today were used in the *Principia* but not named. The mathematical aspects of the first two books were so clearly consistent that they were easily accepted; for example, Locke asked Huygens whether he could trust the mathematical proofs, and was assured about their correctness.

However, the concept of an attractive force acting at a distance received a cooler response. In his notes, Newton wrote that the inverse square law arose naturally due to the structure of matter. However, he retracted this sentence in the published version, where he stated that the motion of planets is consistent with an inverse square law, but refused to speculate on the origin of the law. Huygens and Leibniz noted that the law was incompatible with the notion of the aether. From a Cartesian point of view, therefore, this was a faulty theory. Newton's defence has been adopted since by many famous physicists—he pointed out that the mathematical form of the theory had to be correct since it explained the data, and he refused to speculate further on the basic nature of gravity. The sheer number of phenomena that could be organised by the theory was so impressive that younger "philosophers" soon adopted the methods and language of the *Principia*.

Rules of Reasoning in Philosophy

Perhaps to reduce the risk of public misunderstanding, Newton included at the beginning of Book 3 (in the second (1713) and third (1726) editions) a section entitled "Rules of Reasoning in Philosophy." In the four rules, as they came finally to stand in the 1726 edition, Newton effectively offers a methodology for handling unknown phenomena in nature and reaching towards explanations for them. The four Rules of the 1726 edition run as follows (omitting some explanatory comments that follow each):

Rule 1: *We are to admit no more causes of natural things than such as are both true and sufficient to explain their appearances.*

Rule 2: Therefore to the same natural effects we must, as far as possible, assign the same causes.

Rule 3: The qualities of bodies, which admit neither intensification nor remission of degrees, and which are found to belong to all bodies within the reach of our experiments, are to be esteemed the universal qualities of all bodies whatsoever.

Rule 4: In experimental philosophy we are to look upon propositions inferred by general induction from phenomena as accurately or very nearly true, not withstanding any contrary hypothesis that may be imagined, till such time as other phenomena occur, by which they may either be made more accurate, or liable to exceptions.

This section of Rules for philosophy is followed by a listing of 'Phenomena', in which are listed a number of mainly astronomical observations, that Newton used as the basis for inferences later on, as if adopting a consensus set of facts from the astronomers of his time.

Both the 'Rules' and the 'Phenomena' evolved from one edition of the *Principia* to the next. Rule 4 made its appearance in the third (1726) edition; Rules 1–3 were present as 'Rules' in the second (1713) edition, and predecessors of them were also present in the first edition of 1687, but there they had a different heading: they were not given as 'Rules', but rather in the first (1687) edition the predecessors of the three later 'Rules', and of most of the later 'Phenomena', were all lumped together under a single heading 'Hypotheses' (in which the third item was the predecessor of a heavy revision that gave the later Rule 3).

From this textual evolution, it appears that Newton wanted by the later headings 'Rules' and 'Phenomena' to clarify for his readers his view of the roles to be played by these various statements.

In the third (1726) edition of the *Principia*, Newton explains each rule in an alternative way and/or gives an example to back up what the rule is claiming. The first rule is explained as a philosophers' principle of economy. The second rule states that if one cause is assigned to a natural effect, then the same cause so far as possible must be assigned to natural effects of the same kind: for example respiration in humans and in animals, fires in the home and in the Sun, or the reflection of light whether it occurs terrestrially or from the planets. An extensive explanation is given of the third rule, concerning the qualities of bodies, and Newton discusses here the generalisation of observational results, with a caution against making up fancies contrary to experiments, and use of the rules to illustrate the observation of gravity and space.

Isaac Newton's statement of the four rules revolutionised the investigation of phenomena. With these rules, Newton could in principle begin to address all of the world's present unsolved mysteries. He was able to use his new analytical method to replace that of Aristotle, and he was able to use his method to tweak and update Galileo's experimental method. The re-creation of Galileo's method has never been significantly changed and in its substance, scientists use it today.

General Scholium

The *General Scholium* is a concluding essay added to the second edition, 1713 (and amended in the third edition, 1726). It is not to be confused with the *General Scholium* at the end of Book 2, Section 6, which discusses his pendulum experiments and resistance due to air, water, and other fluids.

Here Newton used what became his famous expression Hypotheses non fingo, "I formulate no hypotheses", in response to criticisms of the first edition of the *Principia*. (*'Fingo'* is sometimes nowadays translated 'feign' rather than the traditional 'frame'.) Newton's gravitational attraction, an invisible force able to act over vast distances, had led to criticism that he had introduced "occult agencies" into science. Newton firmly rejected such criticisms and wrote that it was enough that the phenomena implied gravitational attraction, as they did; but the phenomena did not so far indicate the cause of this gravity, and it was both unnecessary and improper to frame hypotheses of things not implied by the phenomena: such hypotheses "have no place in experimental philosophy", in contrast to the proper way in which "particular propositions are inferr'd from the phenomena and afterwards rendered general by induction".

Newton also underlined his criticism of the vortex theory of planetary motions, of Descartes, pointing to its incompatibility with the highly eccentric orbits of comets, which carry them "through all parts of the heavens indifferently".

Newton also gave theological argument. From the system of the world, he inferred the existence of a Lord God, along lines similar to what is sometimes called the argument from intelligent or purposive design. It has been suggested that Newton gave "an oblique argument for a unitarian conception of God and an implicit attack on the doctrine of the Trinity", but the General Scholium appears to say nothing specifically about these matters.

Halley and Newton's Initial Stimulus

In January 1684, Edmond Halley, Christopher Wren and Robert Hooke had a conversation in which Hooke claimed to not only have derived the inverse-square law, but also all the laws of planetary motion. Wren was unconvinced, Hooke did not produce the claimed derivation although the others gave him time to do it, and Halley, who could derive the inverse-square law for the restricted circular case (by substituting Kepler's relation into Huygens' formula for the centrifugal force) but failed to derive the relation generally, resolved to ask Newton.

Halley's visits to Newton in 1684 thus resulted from Halley's debates about planetary motion with Wren and Hooke, and they seem to have provided Newton with the incentive and spur to develop and write what became *Philosophiae Naturalis Principia Mathematica* (*Mathematical Principles of Natural Philosophy*). Halley was at that time a Fellow and Council member of the Royal Society in London (positions that in 1686 he resigned to become the Society's paid Clerk). Halley's visit to Newton in Cambridge in 1684 probably occurred in August. When Halley asked Newton's opinion on the problem of planetary motions discussed earlier that year between Halley, Hooke and Wren, Newton surprised Halley by saying that he had already made the derivations some time ago; but that he could not find the papers. (Matching accounts of this meeting come from Halley and Abraham De Moivre to whom Newton confided.) Halley then had to wait for Newton to 'find' the results, but in November 1684 Newton sent Halley an amplified version of whatever previous work Newton had done on the subject. This took the form of a 9-page manuscript, *De motu corporum in gyrum* (*Of the motion of bodies in an orbit*): the title is shown on some surviving copies, although the (lost) original may have been without title.

Newton's tract *De motu corporum in gyrum*, which he sent to Halley in late 1684, derived what are now known as the three laws of Kepler, assuming an inverse square law of force, and generalised the result to conic sections. It also extended the methodology by adding the solution of a problem on the motion of a body through a resisting medium. The contents of *De motu* so excited Halley by their mathematical and physical originality and far-reaching implications for astronomical theory, that he immediately went to visit Newton again, in November 1684, to ask Newton to let the Royal Society have more of such work. The results of their meetings clearly helped to stimulate Newton with the enthusiasm needed to take his investigations of mathematical problems much further in this area of physical science, and he did so in a period of highly concentrated work that lasted at least until mid-1686.

Newton's single-minded attention to his work generally, and to his project during this time, is shown by later reminiscences from his secretary and copyist of the period, Humphrey Newton. His account tells of Isaac Newton's absorption in his studies, how he sometimes forgot his food, or his sleep, or the state of his clothes, and how when he took a walk in his garden he would sometimes rush back to his room with some new thought, not even waiting to sit before beginning to write it down. Other evidence also shows Newton's absorption in the *Principia*: Newton for years kept up a regular programme of chemical or alchemical experiments, and he normally kept dated notes of them, but for a period from May 1684 to April 1686, Newton's chemical notebooks have no entries at all. So it seems that Newton abandoned pursuits to which he was normally dedicated, and did very little else for well over a year and a half, but concentrated on developing and writing what became his great work.

The first of the three constituent books was sent to Halley for the printer in spring 1686, and the other two books somewhat later. The complete work, published by Halley at his own financial risk, appeared in July 1687. Newton had also communicated *De motu* to Flamsteed, and during the period of composition he exchanged a few letters with Flamsteed about observational data on the planets, eventually acknowledging Flamsteed's contributions in the published version of the *Principia* of 1687.

Preliminary Version

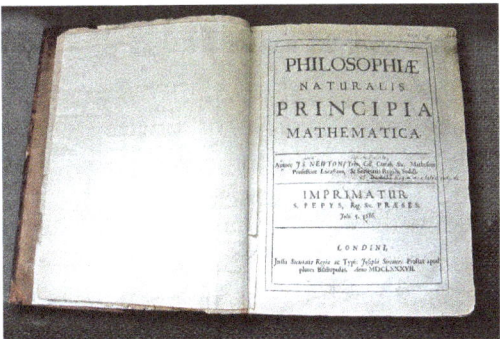

Newton's own first edition copy of his *Principia*, with handwritten corrections for the second edition.

The process of writing that first edition of the *Principia* went through several stages and drafts: some parts of the preliminary materials still survive, while others are lost except for fragments and cross-references in other documents.

Surviving materials show that Newton (up to some time in 1685) conceived his book as a two-volume work. The first volume was to be titled *De motu corporum, Liber primus*, with contents that later appeared in extended form as Book 1 of the *Principia*.

A fair-copy draft of Newton's planned second volume *De motu corporum, Liber secundus* survives, its completion dated to about the summer of 1685. It covers the application of the results of *Liber primus* to the Earth, the Moon, the tides, the solar system, and the universe; in this respect it has much the same purpose as the final Book 3 of the *Principia*, but it is written much less formally and is more easily read.

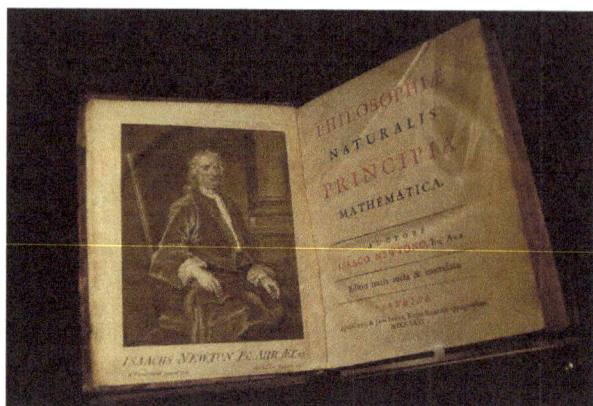

Titlepage and frontispiece of the third edition, London, 1726 (John Rylands Library)

It is not known just why Newton changed his mind so radically about the final form of what had been a readable narrative in *De motu corporum, Liber secundus* of 1685, but he largely started afresh in a new, tighter, and less accessible mathematical style, eventually to produce Book 3 of the *Principia* as we know it. Newton frankly admitted that this change of style was deliberate when he wrote that he had (first) composed this book "in a popular method, that it might be read by many", but to "prevent the disputes" by readers who could not "lay aside the[ir] prejudices", he had "reduced" it "into the form of propositions (in the mathematical way) which should be read by those only, who had first made themselves masters of the principles established in the preceding books". The final Book 3 also contained in addition some further important quantitative results arrived at by Newton in the meantime, especially about the theory of the motions of comets, and some of the perturbations of the motions of the Moon.

The result was numbered Book 3 of the *Principia* rather than Book 2, because in the meantime, drafts of *Liber primus* had expanded and Newton had divided it into two books. The new and final Book 2 was concerned largely with the motions of bodies through resisting mediums.

But the *Liber secundus* of 1685 can still be read today. Even after it was superseded by Book 3 of the *Principia*, it survived complete, in more than one manuscript. After Newton's death in 1727, the relatively accessible character of its writing encouraged the publication of an English translation in 1728 (by persons still unknown, not authorised by Newton's heirs). It appeared under the English title *A Treatise of the System of the World*. This had some amendments relative to Newton's manuscript of 1685, mostly to remove cross-references that used obsolete numbering to cite the propositions of an early draft of Book 1 of the *Principia*. Newton's heirs shortly afterwards published the Latin version in their possession, also in 1728, under the (new) title *De Mundi Systemate*, amended to update cross-references, citations and diagrams to those of the later editions of the *Principia*, making it look superficially as if it had been written by Newton after the *Principia*, rather than before. The *System of the World* was sufficiently popular to stimulate two revisions (with similar changes as in the Latin printing), a second edition (1731), and a 'corrected' reprint of the second edition (1740).

Halley's Role as Publisher

The text of the first of the three books of the *Principia* was presented to the Royal Society at the close of April 1686. Hooke made some priority claims (but failed to substantiate them), causing some delay. When Hooke's claim was made known to Newton, who hated disputes, Newton threatened to withdraw and suppress Book 3 altogether, but Halley, showing considerable diplomatic skills, tactfully persuaded Newton to withdraw his threat and let it go forward to publication. Samuel Pepys, as President, gave his imprimatur on 30 June 1686, licensing the book for publication. The Society had just spent its book budget on a *History of Fishes*, and the cost of publication was borne by Edmund Halley (who was also then acting as publisher of the *Philosophical Transactions of the Royal Society*): the book appeared in summer 1687.

Historical Context

Beginnings of the Scientific Revolution

Nicolaus Copernicus (1473–1543) was the first person to formulate a
comprehensive heliocentric (or Sun-centered) model of the universe

Nicolaus Copernicus had moved the Earth away from the center of the universe with the heliocentric theory for which he presented evidence in his book *De revolutionibus orbium coelestium* (*On the revolutions of the heavenly spheres*) published in 1543. The structure was completed when Johannes Kepler wrote the book *Astronomia nova* (*A new astronomy*) in 1609, setting out the evidence that planets move in elliptical orbits with the sun at one focus, and that planets do not move with constant speed along this orbit. Rather, their speed varies so that the line joining the centres of the sun and a planet sweeps out equal areas in equal times. To these two laws he added a third a decade later, in his book *Harmonices Mundi* (*Harmonies of the world*). This law sets out a proportionality between the third power of the characteristic distance of a planet from the sun and the square of the length of its year.

The foundation of modern dynamics was set out in Galileo's book *Dialogo sopra i due massimi sistemi del mondo* (*Dialogue on the two main world systems*) where the notion of inertia was implicit and used. In addition, Galileo's experiments with inclined planes had yielded precise mathematical relations between elapsed time and acceleration, velocity or distance for uniform and uniformly accelerated motion of bodies.

Descartes' book of 1644 *Principia philosophiae* (*Principles of philosophy*) stated that bodies can act on each other only through contact: a principle that induced people, among them himself, to hypothesize a universal medium as the carrier of interactions such as light and gravity—the aether. Newton was criticized for apparently introducing forces that acted at distance without any medium. Not until the development of particle theory was Descartes' notion vindicated when it was possible to describe all interactions, like the strong, weak, and electromagnetic fundamental interactions, using mediating gauge bosons and gravity through hypothesized gravitons. Although he was mistaken in his treatment of circular motion, this effort was more fruitful in the short term when it led others to identify circular motion as a problem raised by the principle of inertia. Christiaan Huygens solved this problem in the 1650s and published it much later in 1673 in his book *Horologium oscillatorium sive de motu pendulorum*.

Newton's Role

Newton had studied these books, or, in some cases, secondary sources based on them, and taken notes entitled *Quaestiones quaedam philosophicae* (*Questions about philosophy*) during his days as an undergraduate. During this period (1664–1666) he created the basis of calculus, and performed the first experiments in the optics of colour. At this time, his proof that white light was a combination of primary colours (found via prismatics) replaced the prevailing theory of colours and received an overwhelmingly favourable response, and occasioned bitter disputes with Robert Hooke and others, which forced him to sharpen his ideas to the point where he already composed sections of his later book *Opticks* by the 1670s in response. Work on calculus is shown in various papers and letters, including two to Leibniz. He became a fellow of the Royal Society and the second Lucasian Professor of Mathematics (succeeding Isaac Barrow) at Trinity College, Cambridge.

Newton's Early Work on Motion

In the 1660s Newton studied the motion of colliding bodies, and deduced that the centre of mass of two colliding bodies remains in uniform motion. Surviving manuscripts of the 1660s also show Newton's interest in planetary motion and that by 1669 he had shown, for a circular case of planetary motion, that the force he called 'endeavour to recede' (now called centrifugal force) had an inverse-square relation with distance from the center. After his 1679–1680 correspondence with Hooke, described below, Newton adopted the language of inward or centripetal force. According to Newton scholar J Bruce Brackenridge, although much has been made of the change in language and difference of point of view, as between centrifugal or centripetal forces, the actual computations and proofs remained the same either way. They also involved the combination of tangential and radial displacements, which Newton was making in the 1660s. The difference between the centrifugal and centripetal points of view, though a significant change of perspective, did not change the analysis. Newton also clearly expressed the concept of linear inertia in the 1660s: for this Newton was indebted to Descartes' work published 1644.

Controversy with Hooke

Hooke published his ideas about gravitation in the 1660s and again in 1674. He argued for an attracting principle of gravitation in *Micrographia* of 1665, in a 1666 Royal Society lecture *On gravity*, and again in 1674, when he published his ideas about the *System of the World* in somewhat developed form, as an addition to *An Attempt to Prove the Motion of the Earth from Observations*. Hooke clearly postulated mutual attractions between the Sun and planets, in a way that increased with nearness to the attracting body, along with a principle of linear inertia. Hooke's statements up to 1674 made no mention, however, that an inverse square law applies or might apply to these attractions. Hooke's gravitation was also not yet universal, though it approached universality more closely than previous hypotheses. Hooke also did not provide accompanying evidence or mathematical demonstration. On these two aspects, Hooke stated in 1674: "Now what these several degrees [of gravitational attraction] are I have not yet experimentally verified" (indicating that he did not yet know what law the gravitation might follow); and as to his whole proposal: "This I only hint at present", "having my self many other things in hand which I would first compleat, and therefore cannot so well attend it" (i.e., "prosecuting this Inquiry").

Imaginary portrait of English polymath Robert Hooke (1635–1703).

In November 1679, Hooke began an exchange of letters with Newton, of which the full text is now published. Hooke told Newton that Hooke had been appointed to manage the Royal Society's correspondence, and wished to hear from members about their researches, or their views about the researches of others; and as if to whet Newton's interest, he asked what Newton thought about various matters, giving a whole list, mentioning "compounding the celestial motions of the planets of a direct motion by the tangent and an attractive motion towards the central body", and "my hypothesis of the lawes or causes of springinesse", and then a new hypothesis from Paris about planetary motions (which Hooke described at length), and then efforts to carry out or improve national surveys, the difference of latitude between London and Cambridge, and other items. Newton's reply offered "a fansy of my own" about a terrestrial experiment (not a proposal about celestial motions) which might detect the Earth's motion, by the use of a body first suspended in air and then dropped to let it fall. The main point was to indicate how Newton thought the falling body could experimentally reveal the Earth's motion by its direction of deviation from the vertical, but he went on hypothetically to consider how its motion could continue if the solid Earth had not been in the way (on a spiral path to the centre). Hooke disagreed with Newton's idea of how the body would continue to move. A short further correspondence developed, and towards the end of it Hooke, writing on 6 January 1680 to Newton, communicated his "supposition ... that the Attraction always is in a duplicate proportion to the Distance from the Center Reciprocall, and Consequently that the Velocity will be in a subduplicate proportion to the Attraction and Consequently as Kepler Supposes Reciprocall to the Distance." (Hooke's inference about the velocity was actually incorrect.)

In 1686, when the first book of Newton's *Principia* was presented to the Royal Society, Hooke claimed that Newton had obtained from him the "notion" of "the rule of the decrease of Gravity, being reciprocally as the squares of the distances from the Center". At the same time (according to Edmond Halley's contemporary report) Hooke agreed that "the Demonstration of the Curves generated therby" was wholly Newton's.

A recent assessment about the early history of the inverse square law is that "by the late 1660s," the assumption of an "inverse proportion between gravity and the square of distance was rather common and had been advanced by a number of different people for different reasons". Newton himself had shown in the 1660s that for planetary motion under a circular assumption, force in the radial direction had an inverse-square relation with distance from the center. Newton, faced in May 1686 with Hooke's claim on the inverse square law, denied that Hooke was to be credited as author of the idea, giving reasons including the citation of prior work by others before Hooke. Newton also firmly claimed that even if it had happened that he had first heard of the inverse square proportion from Hooke, which it had not, he would still have some rights to it in view of his mathematical developments and demonstrations, which enabled observations to be relied on as evidence of its accuracy, while Hooke, without mathematical demonstrations and evidence in favour of the supposition, could only guess (according to Newton) that it was approximately valid "at great distances from the center".

The background described above shows there was basis for Newton to deny deriving the inverse square law from Hooke. On the other hand, Newton did accept and acknowledge, in all editions of the *Principia*, that Hooke (but not exclusively Hooke) had separately appreciated the inverse square law in the solar system. Newton acknowledged Wren, Hooke and Halley in this connection in the Scholium to Proposition 4 in Book 1. Newton also acknowledged to Halley that his correspondence with Hooke in 1679–80 had reawakened his dormant interest in astronomical matters, but that did not mean, according to Newton, that Hooke had told Newton anything new or original: "yet am I not beholden to him for any light into that business but only for the diversion he gave me from my other studies to think on these things & for his dogmaticalness in writing as if he had found the motion in the Ellipsis, which inclined me to try it ...".) Newton's reawakening interest in astronomy received further stimulus by the appearance of a comet in the winter of 1680/1681, on which he corresponded with John Flamsteed.

In 1759, decades after the deaths of both Newton and Hooke, Alexis Clairaut, mathematical astronomer eminent in his own right in the field of gravitational studies, made his assessment after reviewing what Hooke had published on gravitation. "One must not think that this idea ... of Hooke diminishes Newton's glory", Clairaut wrote; "The example of Hooke" serves "to show what a distance there is between a truth that is glimpsed and a truth that is demonstrated".

Location of Early-edition Copies

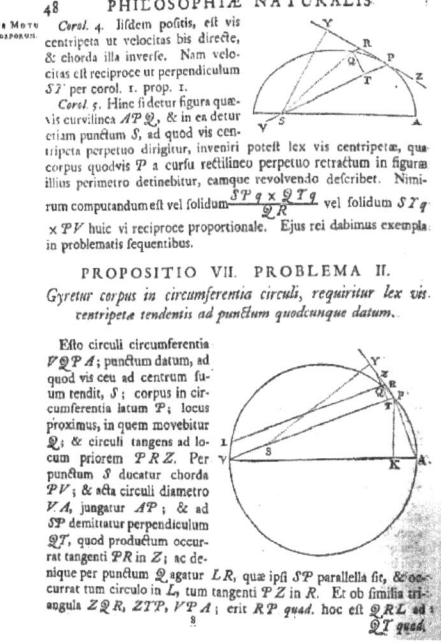

A page from the *Principia*

Since only between 250 and 400 copies were printed by the Royal Society, the first edition is very rare. Several rare-book collections contain first edition and other early copies of Newton's *Principia Mathematica*, including:

- Cambridge University Library has Newton's own copy of the first edition, with handwritten notes for the second edition.

- The Earl Gregg Swem Library at the College of William & Mary has a first edition copy of the Principia. In it are notes in Latin throughout by a not yet identified hand.

- The Frederick E. Brasch Collection of Newton and Newtoniana in Stanford University also has a first edition of the Principia.

- A first edition forms part of the Crawford Collection, housed at the Royal Observatory, Edinburgh.

- The Uppsala University Library owns a first edition copy, which was stolen in the 1960s and returned to the library in 2009.

- The Folger Shakespeare Library in Washington, D.C. owns a first edition, as well as a 1713 second edition.

- The Huntington Library in San Marino, California owns Isaac Newton's personal copy, with annotations in Newton's own hand.

- The Martin Bodmer Library keeps a copy of the original edition that was owned by Leibniz. In it, we can see handwritten notes by Leibniz, in particular concerning the controversy of who first formulated calculus (although he published it later, Newton argued that he developed it earlier).

In 2016, a first edition sold for $3.7 million.

A facsimile edition (based on the 3rd edition of 1726 but with variant readings from earlier editions and important annotations) was published in 1972 by Alexandre Koyré and I. Bernard Cohen.

Later Editions

Two later editions were published by Newton:

Second Edition, 1713

Newton had been urged to make a new edition of the *Principia* since the early 1690s, partly because copies of the first edition had already become very rare and expensive within a few years after 1687. Newton referred to his plans for a second edition in correspondence with Flamsteed in November 1694: Newton also maintained annotated copies of the first edition specially bound up with interleaves on which he could note his revisions; two of these copies still survive: but he had not completed the revisions by 1708, and of two would-be editors, Newton had almost severed connections with one, Nicolas Fatio de Duillier, and the other, David Gregory seems not to have met with Newton's approval and was also terminally ill, dying later in 1708. Nevertheless, reasons were accumulating not to put off the new edition any longer. Richard Bentley, master of Trinity College, persuaded Newton to allow him to undertake a second edition, and in June 1708 Bentley wrote to Newton with a specimen print of the first sheet, at the same time expressing the (unfulfilled) hope that Newton had made progress towards finishing the revisions. It seems that Bentley then realised that the editorship was technically too difficult for him, and with Newton's consent he appointed Roger Cotes, Plumian professor of astronomy at Trinity, to undertake the editorship for him as a kind of deputy (but Bentley still made the publishing arrangements and had the financial responsibility and profit). The correspondence of 1709–1713 shows Cotes reporting to two masters, Bentley and Newton, and managing (and often correcting) a large and important set of revisions to which Newton sometimes could not give his full attention. Under the weight of Cotes' efforts, but impeded by priority disputes between Newton and Leibniz, and by troubles at the Mint, Cotes was able to announce publication to Newton on 30 June 1713. Bentley sent Newton only six presentation copies; Cotes was unpaid; Newton omitted any acknowledgement to Cotes.

Among those who gave Newton corrections for the Second Edition were: Firmin Abauzit, Roger Cotes and David Gregory. However, Newton omitted acknowledgements to some because of the priority disputes. John Flamsteed, the Astronomer Royal, suffered this especially.

Third Edition, 1726

The third edition was published 25 March 1726, under the stewardship of *Henry Pemberton, M.D., a man of the greatest skill in these matters...*; Pemberton later said that this recognition was worth more to him than the two hundred guinea award from Newton.

Annotated and Other Editions

In 1739–42, two French priests, Pères Thomas LeSeur and François Jacquier (of the Minim order, but sometimes erroneously identified as Jesuits), produced with the assistance of J.-L. Calandrini an extensively annotated version of the *Principia* in the 3rd edition of 1726. Sometimes this is referred to as the *Jesuit edition*: it was much used, and reprinted more than once in Scotland during the 19th century.

Émilie du Châtelet also made a translation of Newton's Principia into French. Unlike LeSeur and Jacquier's edition, hers was a complete translation of Newton's three books and their prefaces. She also included a Commentary section where she fused the three books into a much clearer and easier to understand summary. She included an analytical section where she applied the new mathematics of calculus to Newton's most controversial theories. Previously, geometry was the standard mathematics used to analyse theories. Du Châtelet's translation is the only complete one to have been done in French and hers remains the standard French translation to this day.

English Translations

Two full English translations of Newton's *Principia* have appeared, both based on Newton's 3rd edition of 1726.

The first, from 1729, by Andrew Motte, was described by Newton scholar I. Bernard Cohen (in 1968) as "still of enormous value in conveying to us the sense of Newton's words in their own time, and it is generally faithful to the original: clear, and well written". The 1729 version was the basis for several republications, often incorporating revisions, among them a widely used modernised English version of 1934, which appeared under the editorial name of Florian Cajori (though completed and published only some years after his death). Cohen pointed out ways in which the 18th-century terminology and punctuation of the 1729 translation might be confusing to modern readers, but he also made severe criticisms of the 1934 modernised English version, and showed that the revisions had been made without regard to the original, also demonstrating gross errors "that provided the final impetus to our decision to produce a wholly new translation".

The second full English translation, into modern English, is the work that resulted from this decision by collaborating translators I. Bernard Cohen, Anne Whitman, and Julia Budenz; it was published in 1999 with a guide by way of introduction.

William H. Donahue has published a translation of the work's central argument, published in 1996, along with expansion of included proofs and ample commentary. The book was developed as a textbook for classes at St. John's College and the aim of this translation is to be faithful to the Latin text.

Homages

In 2014, British astronaut Tim Peake named his upcoming mission to the International Space Station *Principia* after the book, in "honour of Britain's greatest scientist". Tim Peake's *Principia* launched on December 15, 2015 aboard Soyuz TMA-19M.

Dark Energy

In physical cosmology and astronomy, dark energy is an unknown form of energy which is hypothesized to permeate all of space, tending to accelerate the expansion of the universe. Dark energy is the most accepted hypothesis to explain the observations since the 1990s indicating that the universe is expanding at an accelerating rate.

Assuming that the standard model of cosmology is correct, the best current measurements indicate that dark energy contributes 68.3% of the total energy in the present-day observable universe. The mass–energy of dark matter and ordinary (baryonic) matter contribute 26.8% and 4.9%, respectively, and other components such as neutrinos and photons contribute a very small amount. The density of dark energy ($\sim 7 \times 10^{-30}$ g/cm^3) is very low, much less than the density of ordinary matter or dark matter within galaxies. However, it comes to dominate the mass–energy of the universe because it is uniform across space.

Two proposed forms for dark energy are the cosmological constant, representing a constant energy density filling space homogeneously, and scalar fields such as quintessence or moduli, dynamic quantities whose energy density can vary in time and space. Contributions from scalar fields that are constant in space are usually also included in the cosmological constant. The cosmological constant can be formulated to be equivalent to the zero-point radiation of space i.e. the vacuum energy. Scalar fields that change in space can be difficult to distinguish from a cosmological constant because the change may be extremely slow.

High-precision measurements of the expansion of the universe are required to understand how the expansion rate changes over time and space. In general relativity, the evolution of the expansion rate is estimated from the curvature of the universe and the cosmological equation of state (the relationship between temperature, pressure, and combined matter, energy, and vacuum energy density for any region of space). Measuring the equation of state for dark energy is one of the biggest efforts in observational cosmology today. Adding the cosmological constant to cosmology's standard FLRW metric leads to the Lambda-CDM model, which has been referred to as the "*standard model of cosmology*" because of its precise agreement with observations.

History of Discovery and Previous Speculation

Einstein's Cosmological Constant

The "cosmological constant" is a constant term that can be added to Einstein's field equation of General Relativity. If considered as a "source term" in the field equation, it can be viewed as equivalent to the mass of empty space (which conceptually could be either positive or negative), or "vacuum energy".

The cosmological constant was first proposed by Einstein as a mechanism to obtain a solution of the gravitational field equation that would lead to a static universe, effectively using dark energy to balance gravity. Einstein gave the consmological constant the symbol Λ (capital lambda).

The mechanism was an example of fine-tuning, and it was later realized that Einstein's static universe would not be stable: local inhomogeneities would ultimately lead to either the runaway expansion or contraction of the universe. The equilibrium is unstable: if the universe expands slightly, then the expansion releases vacuum energy, which causes yet more expansion. Likewise, a universe which contracts slightly will continue contracting. These sorts of disturbances are inevitable, due to the uneven distribution of matter throughout the universe. Further, observations made by Edwin Hubble in 1929 showed that the universe appears to be expanding and not static at all. Einstein reportedly referred to his failure to predict the idea of a dynamic universe, in contrast to a static universe, as his greatest blunder.

Inflationary Dark Energy

Alan Guth and Alexei Starobinsky proposed in 1980 that a negative pressure field, similar in concept to dark energy, could drive cosmic inflation in the very early universe. Inflation postulates that some repulsive force, qualitatively similar to dark energy, resulted in an enormous and exponential expansion of the universe slightly after the Big Bang. Such expansion is an essential feature of most current models of the Big Bang. However, inflation must have occurred at a much higher energy density than the dark energy we observe today and is thought to have completely ended when the universe was just a fraction of a second old. It is unclear what relation, if any, exists between dark energy and inflation. Even after inflationary models became accepted, the cosmological constant was thought to be irrelevant to the current universe.

Nearly all inflation models predict that the total (matter+energy) density of the universe should be very close to the critical density. During the 1980s, most cosmological research focused on models with critical density in matter only, usually 95% cold dark matter and 5% ordinary matter (baryons). These models were found to be successful at forming realistic galaxies and clusters, but some problems appeared in the late 1980s: in particular, the model required a value for the Hubble constant lower than preferred by observations, and the model under-predicted observations of large-scale galaxy clustering. These difficulties became stronger after the discovery of anisotropy in the cosmic microwave background by the COBE spacecraft in 1992, and several modified CDM models came under active study through the mid-1990s: these included the Lambda-CDM model and a mixed cold/hot dark matter model. The first direct evidence for dark energy came from supernova observations in 1998 of accelerated expansion in Riess *et al.* and in Perlmutter *et al.*, and the Lambda-CDM model then became the leading model. Soon after, dark energy was supported by independent observations: in 2000, the BOOMERanG and Maxima cosmic microwave background experiments observed the first acoustic peak in the CMB, showing that the total (matter+energy) density is close to 100% of critical density. Then in 2001, the 2dF Galaxy Redshift Survey gave strong evidence that the matter density is around 30% of critical. The large difference between these two supports a smooth component of dark energy making up the difference. Much more precise measurements from WMAP in 2003–2010 have continued to support the standard model and give more accurate measurements of the key parameters.

The term "dark energy", echoing Fritz Zwicky's "dark matter" from the 1930s, was coined by Michael Turner in 1998.

As of 2013, the Lambda-CDM model is consistent with a series of increasingly rigorous cosmological observations, including the Planck spacecraft and the Supernova Legacy Survey. First results from the SNLS reveal that the average behavior (i.e., equation of state) of dark energy behaves like Einstein's cosmological constant to a precision of 10%. Recent results from the Hubble Space

Telescope Higher-Z Team indicate that dark energy has been present for at least 9 billion years and during the period preceding cosmic acceleration.

Nature

The nature of dark energy is more hypothetical than that of dark matter, and many things about the nature of dark energy remain matters of speculation. Dark energy is thought to be very homogeneous, not very dense and is not known to interact through any of the fundamental forces other than gravity. Since it is quite rarefied — roughly 10^{-27} kg/m^3 — it is unlikely to be detectable in laboratory experiments. The reason dark energy can have such a profound effect on the universe, making up 68% of universal density, in spite of being so rarefied is because it uniformly fills otherwise empty space.

Independently of its actual nature, dark energy would need to have a strong negative pressure (acting repulsively) like radiation pressure in a metamaterial to explain the observed acceleration of the expansion of the universe. According to general relativity, the pressure within a substance contributes to its gravitational attraction for other things just as its mass density does. This happens because the physical quantity that causes matter to generate gravitational effects is the stress–energy tensor, which contains both the energy (or matter) density of a substance and its pressure and viscosity. In the Friedmann–Lemaître–Robertson–Walker metric, it can be shown that a strong constant negative pressure in all the universe causes an acceleration in universe expansion if the universe is already expanding, or a deceleration in universe contraction if the universe is already contracting. This accelerating expansion effect is sometimes labeled "gravitational repulsion".

Technical Definition

In standard cosmology, there are three components of the universe: matter, radiation and dark energy. Matter is anything whose energy density scales with the inverse cube of the scale factor, i.e. $\rho \propto a^{-3}$, while radiation is anything which scales to the inverse fourth power of the scale factor $\rho \propto a^{-4}$. This can be understood intuitively: for an ordinary particle in a square box, doubling the length of a side of the box decreases the density (and hence energy density) by a factor of eight (2^3). For radiation, the decrease in energy density is greater, because an increase in spatial distance also causes a redshift.

The final component, dark energy, is an intrinsic property of space, and so has a constant energy density regardless of the volume under consideration ($\rho \propto a^0$).

Evidence of Existence

The evidence for dark energy is indirect but comes from three independent sources:

- Distance measurements and their relation to redshift, which suggest the universe has expanded more in the last half of its life.

- The theoretical need for a type of additional energy that is not matter or dark matter to form the observationally flat universe (absence of any detectable global curvature).

- It can be inferred from measures of large scale wave-patterns of mass density in the universe.

Supernovae

A Type Ia supernova (bright spot on the bottom-left) near a galaxy

In 1998, the High-Z Supernova Search Team published observations of Type Ia ("one-A") supernovae. In 1999, the Supernova Cosmology Project followed by suggesting that the expansion of the universe is accelerating. The 2011 Nobel Prize in Physics was awarded to Saul Perlmutter, Brian P. Schmidt and Adam G. Riess for their leadership in the discovery.

Since then, these observations have been corroborated by several independent sources. Measurements of the cosmic microwave background, gravitational lensing, and the large-scale structure of the cosmos as well as improved measurements of supernovae have been consistent with the Lambda-CDM model. Some people argue that the only indication for the existence of dark energy is observations of distance measurements and the associated redshifts. Cosmic microwave background anisotropies and baryon acoustic oscillations only serve to demonstrate that distances to a given redshift are larger than would be expected from a "dusty" Friedmann–Lemaître universe and the local measured Hubble constant.

Supernovae are useful for cosmology because they are excellent standard candles across cosmological distances. They allow the expansion history of the universe to be measured by looking at the relationship between the distance to an object and its redshift, which gives how fast it is receding from us. The relationship is roughly linear, according to Hubble's law. It is relatively easy to measure redshift, but finding the distance to an object is more difficult. Usually, astronomers use standard candles: objects for which the intrinsic brightness, the absolute magnitude, is known. This allows the object's distance to be measured from its actual observed brightness, or apparent magnitude. Type Ia supernovae are the best-known standard candles across cosmological distances because of their extreme and consistent luminosity.

Recent observations of supernovae are consistent with a universe made up 71.3% of dark energy and 27.4% of a combination of dark matter and baryonic matter.

Cosmic Microwave Background

The existence of dark energy, in whatever form, is needed to reconcile the measured geometry of space with the total amount of matter in the universe. Measurements of cosmic microwave background (CMB) anisotropies indicate that the universe is close to flat. For the shape of the universe to be flat, the mass/energy density of the universe must be equal to the critical density. The total amount of matter in the universe (including baryons and dark matter), as measured from the CMB spectrum, accounts for only about 30% of the critical density. This implies the existence of an additional form of energy to account for the remaining 70%. The Wilkinson Microwave Anisotropy Probe (WMAP) spacecraft seven-year analysis estimated a universe made up of 72.8% dark energy, 22.7% dark matter and 4.5% ordinary matter. Work done in 2013 based on the Planck spacecraft observations of the CMB gave a more accurate estimate of 68.3% of dark energy, 26.8% of dark matter and 4.9% of ordinary matter.

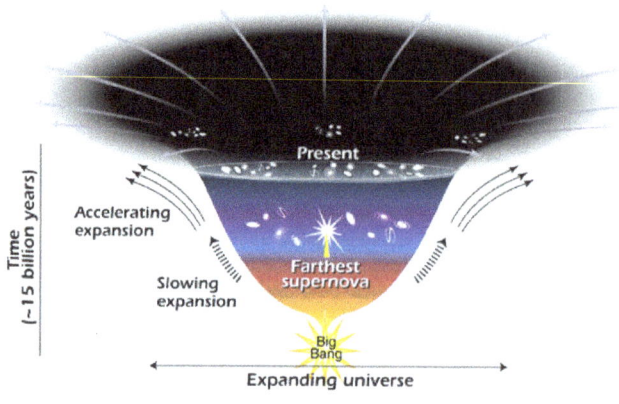

This diagram reveals changes in the rate of expansion since the universe's birth 15 billion years ago. The more shallow the curve, the faster the rate of expansion. The curve changes noticeably about 7.5 billion years ago, when objects in the universe began flying apart at a faster rate. Astronomers theorize that the faster expansion rate is due to a mysterious, dark force that is pushing galaxies apart.

Diagram representing the accelerated expansion of the universe due to dark energy.

Large-scale Structure

The theory of large-scale structure, which governs the formation of structures in the universe (stars, quasars, galaxies and galaxy groups and clusters), also suggests that the density of matter in the universe is only 30% of the critical density.

A 2011 survey, the WiggleZ galaxy survey of more than 200,000 galaxies, provided further evidence towards the existence of dark energy, although the exact physics behind it remains unknown. The WiggleZ survey from the Australian Astronomical Observatory scanned the galaxies to determine their redshift. Then, by exploiting the fact that baryon acoustic oscillations have left voids regularly of ~150 Mpc diameter, surrounded by the galaxies, the voids were used as standard rulers to estimate distances to galaxies as far as 2,000 Mpc (redshift 0.6), allowing for accurate estimate of the speeds of galaxies from their redshift and distance. The data confirmed cosmic acceleration up to half of the age of the universe (7 billion years) and constrain its inhomogeneity to 1 part in 10. This provides a confirmation to cosmic acceleration independent of supernovae.

Late-time Integrated Sachs-Wolfe Effect

Accelerated cosmic expansion causes gravitational potential wells and hills to flatten as photons pass through them, producing cold spots and hot spots on the CMB aligned with vast supervoids and superclusters. This so-called late-time Integrated Sachs–Wolfe effect (ISW) is a direct signal of dark energy in a flat universe. It was reported at high significance in 2008 by Ho *et al.* and Giannantonio *et al.*

Observational Hubble Constant Data

A new approach to test evidence of dark energy through observational Hubble constant data (OHD) has gained significant attention in recent years. The Hubble constant, $H(z)$, is measured as a function of cosmological redshift. OHD directly tracks the expansion history of the universe by taking passively evolving early-type galaxies as "cosmic chronometers". From this point, this approach provides standard clocks in the universe. The core of this idea is the measurement of the differential age evolution as a function of redshift of these cosmic chronometers. Thus, it provides a direct estimate of the Hubble parameter

$$H(z) = -\frac{1}{1+z}\frac{dz}{dt} \approx -\frac{1}{1+z}\frac{\Delta z}{\Delta t}.$$

The reliance on a differential quantity, $\frac{\Delta z}{\Delta t}$, can minimize many common issues and systematic effects; and as a direct measurement of the Hubble parameter instead of its integral, like supernovae and baryon acoustic oscillations (BAO), it brings more information and is appealing in computation. For these reasons, it has been widely used to examine the accelerated cosmic expansion and study properties of dark energy.

Theories of Dark Energy

Dark energy's status as a hypothetical force with unknown properties makes it a very active target of research. The problem is attacked from a great variety of angles, such as modifying the prevailing theory of gravity (general relativity), attempting to pin down the properties of dark energy, and finding alternative ways to explain the observational data.

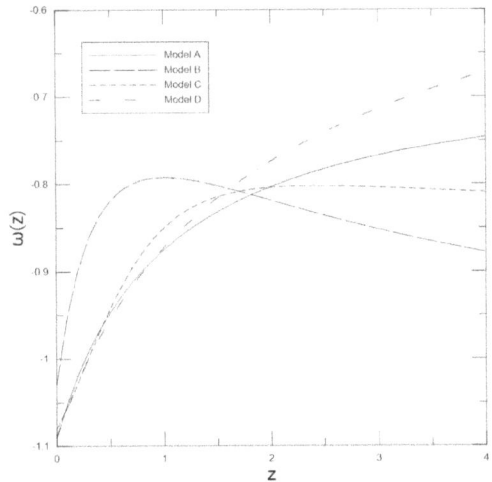

The equation of state of Dark Energy for 4 common models by Redshift.
A: CPL Model,
B: Jassal Model,
C: Barboza & Alcaniz Model,
D: Wetterich Model

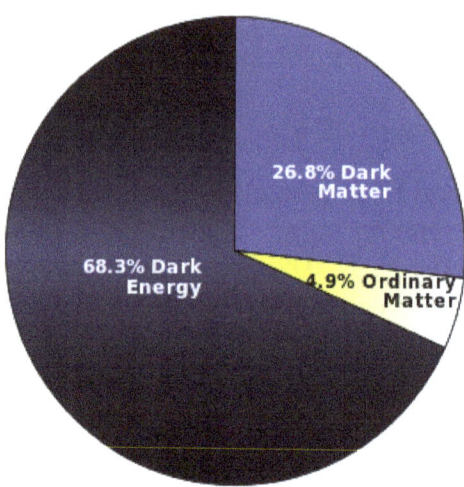

Estimated distribution of matter and energy in the universe

The simplest explanation for dark energy is that it is an intrinsic, fundamental energy of space. This is the cosmological constant, usually represented by the Greek letter Λ (Lambda, hence Lambda-CDM model). Since energy and mass are related according to the equation $E = mc^2$, Einstein's theory of general relativity predicts that this energy will have a gravitational effect. It is sometimes called a vacuum energy because it is the energy density of empty vacuum.

The cosmological constant has negative pressure equal to its energy density and so causes the expansion of the universe to accelerate. The reason a cosmological constant has negative pressure can be seen from classical thermodynamics. In general, energy must be lost from inside a container (the container must do work on its environment) in order for the volume to increase. Specifically, a change in volume dV requires work done equal to a change of energy $-P\,dV$, where P is the pressure. But the amount of energy in a container full of vacuum actually increases when the volume increases, because the energy is equal to ρV, where ρ is the energy density of the cosmological constant. Therefore, P is negative and, in fact, $P = -\rho$.

There are two major advantages for the cosmological constant. The first is that it is simple. Einstein had in fact introduced this term in his original formulation of general relativity such as to get a static universe. Although he later discarded the term after Hubble found that the universe is expanding, a nonzero cosmological constant can act as dark energy, without otherwise changing the Einstein field equations. The other advantage is that there is a natural explanation for its origin. Most quantum field theories predict vacuum fluctuations that would give the vacuum this sort of energy. This is related to the Casimir effect, in which there is a small suction into regions where virtual particles are geometrically inhibited from forming (e.g. between plates with tiny separation).

A major outstanding problem is that the same quantum field theories predict a huge cosmological constant, more than 100 orders of magnitude too large. This would need to be almost, but not exactly, cancelled by an equally large term of the opposite sign. Some supersymmetric theories require a cosmological constant that is exactly zero, which does not help because supersymmetry must be broken.

Nonetheless, the cosmological constant is the most economical solution to the problem of cosmic acceleration. Thus, the current standard model of cosmology, the Lambda-CDM model, includes the cosmological constant as an essential feature.

Modified Gravity

The evidence for dark energy is heavily dependent on the theory of general relativity. Therefore, it is conceivable that a modification to general relativity also eliminates the need for dark energy. There are very many such theories, and research is ongoing.

Quintessence

In quintessence models of dark energy, the observed acceleration of the scale factor is caused by the potential energy of a dynamical field, referred to as quintessence field. Quintessence differs from the cosmological constant in that it can vary in space and time. In order for it not to clump and form structure like matter, the field must be very light so that it has a large Compton wavelength.

No evidence of quintessence is yet available, but it has not been ruled out either. It generally predicts a slightly slower acceleration of the expansion of the universe than the cosmological constant. Some scientists think that the best evidence for quintessence would come from violations of Einstein's equivalence principle and variation of the fundamental constants in space or time. Scalar fields are predicted by the Standard Model of particle physics and string theory, but an analogous problem to the cosmological constant problem (or the problem of constructing models of cosmological inflation) occurs: renormalization theory predicts that scalar fields should acquire large masses.

The coincidence problem asks why the acceleration of the Universe began when it did. If acceleration began earlier in the universe, structures such as galaxies would never have had time to form and life, at least as we know it, would never have had a chance to exist. Proponents of the anthropic principle view this as support for their arguments. However, many models of quintessence have a so-called "tracker" behavior, which solves this problem. In these models, the quintessence field has a density which closely tracks (but is less than) the radiation density until matter-radiation equality, which triggers quintessence to start behaving as dark energy, eventually dominating the universe. This naturally sets the low energy scale of the dark energy.

In 2004, when scientists fit the evolution of dark energy with the cosmological data, they found that the equation of state had possibly crossed the cosmological constant boundary ($w = -1$) from above to below. A No-Go theorem has been proved that gives this scenario at least two degrees of freedom as required for dark energy models. This scenario is so-called Quintom scenario.

Some special cases of quintessence are phantom energy, in which the energy density of quintessence actually increases with time, and k-essence (short for kinetic quintessence) which has a non-standard form of kinetic energy such as a negative kinetic energy. They can have unusual properties: phantom energy, for example, can cause a Big Rip.

Interacting Dark Energy

This class of theories attempts to come up with an all-encompassing theory of both dark matter and dark energy as a single phenomenon that modifies the laws of gravity at various scales. This could for example treat dark energy and dark matter as different facets of the same unknown substance, or postulate that cold dark matter decays into dark energy. Another class of theories that unifies dark matter and dark energy are suggested to be covariant theories of modified gravities. These theories alter the dynamics of the space-time such that the modified dynamic stems what have been assigned to the presence of dark energy and dark matter.

Variable Dark Energy Models

The density of dark energy might have varied in time over the history of the universe. Modern observational data allow for estimates of the present density. Using baryon acoustic oscillations, it is possible to investigate the effect of dark energy in the history of the Universe, and constrain parameters of the equation of state of dark energy. To that end, several models have been proposed. One of the most popular models is the Chevallier–Polarski–Linder model (CPL). Some other common models are, (Barboza & Alcaniz. 2008), (Jassal et al. 2005), (Wetterich. 2004).

Observational Skepticism

Some alternatives to dark energy aim to explain the observational data by a more refined use of established theories. In this scenario, dark energy doesn't actually exist, and is merely a measurement artifact. For example, if we are located in an emptier-than-average region of space, the observed cosmic expansion rate could be mistaken for a variation in time, or acceleration. A different approach uses a cosmological extension of the equivalence principle to show how space might appear to be expanding more rapidly in the voids surrounding our local cluster. While weak, such effects considered cumulatively over billions of years could become significant, creating the illusion of cosmic acceleration, and making it appear as if we live in a Hubble bubble. Yet another possibility is that the accelerated expansion of the universe is an illusion caused by the relative motion of us to the rest of the universe.

Implications for the Fate of the Universe

Cosmologists estimate that the acceleration began roughly 5 billion years ago. Before that, it is thought that the expansion was decelerating, due to the attractive influence of dark matter and baryons. The density of dark matter in an expanding universe decreases more quickly than dark energy, and eventually the dark energy dominates. Specifically, when the volume of the universe doubles, the density of dark matter is halved, but the density of dark energy is nearly unchanged (it is exactly constant in the case of a cosmological constant).

Projections into the future can differ radically for different models of dark energy. For a cosmological constant, or any other model that predicts that the acceleration will continue indefinitely, the ultimate result will be that galaxies outside the Local Group will have a line-of-sight velocity that continually increases with time, eventually far exceeding the speed of light. This is not a violation of special relativity because the notion of "velocity" used here is different from that of velocity in a local inertial frame of reference, which is still constrained to be less than the speed of light for any massive object. Because the Hubble parameter is decreasing with time, there can actually be cases where a galaxy that is receding from us faster than light does manage to emit a signal which reaches us eventually. However, because of the accelerating expansion, it is projected that most galaxies will eventually cross a type of cosmological event horizon where any light they emit past that point will never be able to reach us at any time in the infinite future because the light never reaches a point where its "peculiar velocity" toward us exceeds the expansion velocity away from us (these two notions of velocity are also discussed in Uses of the proper distance). Assuming the dark energy is constant (a cosmological constant), the current distance to this cosmological event horizon is about 16 billion light years, meaning that a signal from an event happening *at present* would eventually be able to reach us in the future if the event were less than 16 billion light years away, but the signal would never reach us if the event were more than 16 billion light years away.

As galaxies approach the point of crossing this cosmological event horizon, the light from them will become more and more redshifted, to the point where the wavelength becomes too large to detect in practice and the galaxies appear to vanish completely. The Earth, the Milky Way, and the Local Group of which the Milky way is a part, would all remain virtually undisturbed as the rest of the universe recedes and disappears from view. In this scenario, the Local Group would ultimately suffer heat death, just as was hypothesized for the flat, matter-dominated universe before measurements of cosmic acceleration.

There are other, more speculative ideas about the future of the universe. The phantom energy model of dark energy results in *divergent* expansion, which would imply that the effective force of dark energy continues growing until it dominates all other forces in the universe. Under this scenario, dark energy would ultimately tear apart all gravitationally bound structures, including galaxies and solar systems, and eventually overcome the electrical and nuclear forces to tear apart atoms themselves, ending the universe in a "Big Rip". It is also possible the universe may never have an end and continue in its present state forever. On the other hand, dark energy might dissipate with time or even become attractive. Such uncertainties leave open the possibility that gravity might yet rule the day and lead to a universe that contracts in on itself in a "Big Crunch", or that there may even be a dark energy cycle, which implies a cyclic model of the universe in which every iteration (Big Bang then eventually a Big Crunch) takes about a trillion (10^{12}) years. While none of these are supported by observations, they are not ruled out.

Cosmological Constant

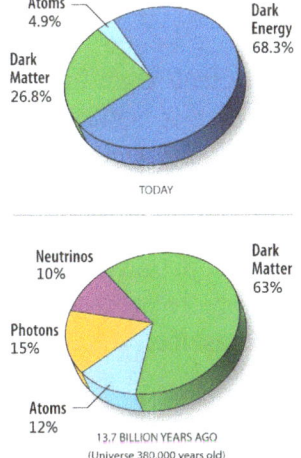

Estimated ratios of dark matter and dark energy (which may be the cosmological constant) in the universe. According to current theories of physics, dark energy now dominates as the largest source of energy of the universe, in contrast to earlier epochs when it was insignificant.

In cosmology, the cosmological constant (usually denoted by the Greek capital letter lambda: Λ) is the value of the energy density of the vacuum of space. It was originally introduced by Albert Einstein in 1917 as an addition to his theory of general relativity to "hold back gravity" and achieve a static universe, which was the accepted view at the time. Einstein abandoned the concept after Hubble's 1929 discovery that all galaxies outside the Local Group (the group that contains the Milky Way Galaxy) are moving away from each other, implying an overall expanding universe. From 1929 until the early 1990s, most cosmology researchers assumed the cosmological constant to be zero.

Since the 1990s, several developments in observational cosmology, especially the discovery of the accelerating universe from distant supernovae in 1998 (in addition to independent evidence from the cosmic microwave background and large galaxy redshift surveys), have shown that around 68% of the mass–energy density of the universe can be attributed to dark energy. While dark energy is poorly understood at a fundamental level, the main required properties of dark energy are that it functions as a type of anti-gravity, it dilutes much more slowly than matter as the universe expands, and it clusters much more weakly than matter, or perhaps not at all. The cosmological constant is the simplest possible form of dark energy since it is constant in both space and time, and this leads to the current standard model of cosmology known as the Lambda-CDM model, which provides a good fit to many cosmological observations.

Equation

The cosmological constant Λ appears in Einstein's field equation in the form of

$$R_{\mu\nu} - \frac{1}{2}R g_{\mu\nu} + \Lambda g_{\mu\nu} = \frac{8\pi G}{c^4} T_{\mu\nu},$$

where R and g describe the structure of spacetime, T pertains to matter and energy affecting that structure, and G and c are conversion factors that arise from using traditional units of measurement. When Λ is zero, this reduces to the original field equation of general relativity. When T is zero, the field equation describes empty space (the vacuum).

The cosmological constant has the same effect as an intrinsic energy density of the vacuum, ρ_{vac} (and an associated pressure). In this context, it is commonly moved onto the right-hand side of the equation, and defined with a proportionality factor of 8π: $\Lambda = 8\pi\rho_{vac}$, where unit conventions of general relativity are used (otherwise factors of G and c would also appear, i.e. $\Lambda = 8\pi(G/c^2)\rho_{vac} = \kappa\rho_{vac}$, where κ is Einstein's constant). It is common to quote values of energy density directly, though still using the name "cosmological constant", with convention $8\pi G = 1$. (The true dimension of Λ is a length^{-2} and it has the value of 1.19×10^{-52} m^{-2} or in reduced Planck units : ~ 3×10^{-122}, calculated with the best present (2015) values of $\Omega_\Lambda = 0.6911\pm0.0062$ and $H_0 = 67.74\pm0.46$ (km/s)/Mpc = $(2.195\pm0.015)\times10^{-18}$ s^{-1}).

A positive vacuum energy density resulting from a cosmological constant implies a negative pressure, and vice versa. If the energy density is positive, the associated negative pressure will drive an accelerated expansion of the universe, as observed.

Ω_Λ (Omega Lambda)

Instead of the cosmological constant itself, cosmologists often refer to the ratio between the energy density due to the cosmological constant and the critical density of the universe, the tipping point for a sufficient density to stop the universe from expanding forever. This ratio is usually denoted Ω_Λ, and is estimated to be 0.6911±0.0062, according to results published by the Planck Collaboration in 2015.

In a flat universe Ω_Λ is the fraction of the energy of the universe due to the cosmological constant, i.e., what we would intuitively call the fraction of the universe that is made up of dark energy. Note that this value changes over time: the critical density changes with cosmological time, but the energy density due to the cosmological constant remains unchanged throughout the history of the universe: the amount of dark energy increases as the universe grows, while the amount of matter does not.

Equation of State

Another ratio that is used by scientists is the equation of state, usually denoted w, which is the ratio of pressure that dark energy puts on the universe to the energy per unit volume. This ratio is $w = -1$ for a true cosmological constant, and is generally different for alternative time-varying forms of vacuum energy such as quintessence.

History

Einstein included the cosmological constant as a term in his field equations for general relativity because he was dissatisfied that otherwise his equations did not allow, apparently, for a static universe: gravity would cause a universe that was initially at dynamic equilibrium to contract. To counteract this possibility, Einstein added the cosmological constant. However, soon after Einstein developed his static theory, observations by Edwin Hubble indicated that the universe appears to be expanding; this was consistent with a cosmological solution to the *original* general relativity equations that had been found by the mathematician Friedmann, working on the Einstein equations of general relativity. Einstein later reputedly referred to his failure to accept the validation of his equations—when they had predicted the expansion of the universe in theory, before it was demonstrated in observation of the cosmological red shift—as the "biggest blunder" of his life.

In fact, adding the cosmological constant to Einstein's equations does not lead to a static universe at equilibrium because the equilibrium is unstable: if the universe expands slightly, then the expansion releases vacuum energy, which causes yet more expansion. Likewise, a universe that contracts slightly will continue contracting.

However, the cosmological constant remained a subject of theoretical and empirical interest. Empirically, the onslaught of cosmological data in the past decades strongly suggests that our universe has a positive cosmological constant. The explanation of this small but positive value is an outstanding theoretical challenge.

Finally, it should be noted that some early generalizations of Einstein's gravitational theory, known as classical unified field theories, either introduced a cosmological constant on theoretical grounds or found that it arose naturally from the mathematics. For example, Sir Arthur

Stanley Eddington claimed that the cosmological constant version of the vacuum field equation expressed the "epistemological" property that the universe is "self-gauging", and Erwin Schrödinger's pure-affine theory using a simple variational principle produced the field equation with a cosmological term.

Positive Value

Observations announced in 1998 of distance–redshift relation for Type Ia supernovae indicated that the expansion of the universe is accelerating. When combined with measurements of the cosmic microwave background radiation these implied a value of $\Omega_\Lambda \approx 0.7$, a result which has been supported and refined by more recent measurements. There are other possible causes of an accelerating universe, such as quintessence, but the cosmological constant is in most respects the simplest solution. Thus, the current standard model of cosmology, the Lambda-CDM model, includes the cosmological constant, which is measured to be on the order of $10^{-52} \, \text{m}^{-2}$, in metric units. Multiplied by other constants that appear in the equations, it is often expressed as $10^{-52} \, \text{m}^{-2}$, $10^{-35} \, \text{s}^{-2}$, $10^{-47} \, \text{GeV}^4$, $10^{-29} \, \text{g/cm}^3$. More recent measurements give the value as $(1.501 \pm 0.043) \times 10^{-25} \, \text{kg/m}^3$. In terms of Planck units, and as a natural dimensionless value, the cosmological constant, Λ, is on the order of 10^{-122}. Modern calculations considering the vacuum energy of all known scalar and vector fields leads to 10^{-54} orders of magnitude smaller than the prediction.

As was only recently seen, by works of 't Hooft, Susskind and others, a positive cosmological constant has surprising consequences, such as a finite maximum entropy of the observable universe.

Predictions

Quantum Field Theory

A major outstanding problem is that most quantum field theories predict a huge value for the quantum vacuum. A common assumption is that the quantum vacuum is equivalent to the cosmological constant. Although no theory exists that supports this assumption, arguments can be made in its favor.

Such arguments are usually based on dimensional analysis and effective field theory. If the universe is described by an effective local quantum field theory down to the Planck scale, then we would expect a cosmological constant of the order of M_{pl}^4. As noted above, the measured cosmological constant is smaller than this by a factor of 10^{-120}. This discrepancy has been called "the worst theoretical prediction in the history of physics!".

Some supersymmetric theories require a cosmological constant that is exactly zero, which further complicates things. This is the *cosmological constant problem*, the worst problem of fine-tuning in physics: there is no known natural way to derive the tiny cosmological constant used in cosmology from particle physics.

Anthropic Principle

One possible explanation for the small but non-zero value was noted by Steven Weinberg in 1987 following the anthropic principle. Weinberg explains that if the vacuum energy took different values in different domains of the universe, then observers would necessarily measure values similar to that which is observed: the formation of life-supporting structures would be suppressed in domains where the vacuum energy is much larger. Specifically, if the vacuum energy is negative and its absolute value is substantially larger than it appears to be in the observed universe (say, a factor of 10 larger), holding all other variables (e.g. matter density) constant, that would mean that the universe is closed; furthermore, its lifetime would be shorter than the age of our universe, possibly too short for intelligent life to form. On the other hand, a universe with a large positive cosmological constant would expand too fast, preventing galaxy formation. According to Weinberg, domains where the vacuum energy is compatible with life would be comparatively rare. Using this argument, Weinberg predicted that the cosmological constant would have a value of less than a hundred times the currently accepted value. In 1992, Weinberg refined this prediction of the cosmological constant to 5 to 10 times the matter density.

This argument depends on a lack of a variation of the distribution (spatial or otherwise) in the vacuum energy density, as would be expected if dark energy were the cosmological constant. There is no evidence that the vacuum energy does vary, but it may be the case if, for example, the vacuum energy is (even in part) the potential of a scalar field such as the residual inflaton. Another theoretical approach that deals with the issue is that of multiverse theories, which predict a large number of "parallel" universes with different laws of physics and/or values of fundamental constants. Again, the anthropic principle states that we can only live in one of the universes that is compatible with some form of intelligent life. Critics claim that these theories, when used as an explanation for fine-tuning, commit the inverse gambler's fallacy.

In 1995, Weinberg's argument was refined by Alexander Vilenkin to predict a value for the cosmological constant that was only ten times the matter density, i.e. about three times the current value since determined.

References

- Vilenkin, Alex (2007). Many worlds in one : the search for other universes. New York: Hill and Wang, A division of Farrar, Straus and Giroux. p. 19. ISBN 978-0-8090-6722-0

- Overbye, Dennis (17 March 2014). "Detection of Waves in Space Buttresses Landmark Theory of Big Bang". New York Times. Retrieved 17 March 2014

- Beringer, J.; et al. (Particle Data Group) (2012). "2013 Review of Particle Physics" (PDF). Phys. Rev. D. 86: 010001. Bibcode:2012PhRvD..86a0001B. doi:10.1103/PhysRevD.86.010001

- Cong Ma; Tongjie Zhang (2010). "Power of Observational Hubble Parameter Data: a Figure of Merit Exploration". Astrophysical Journal. 730 (2): 74. Bibcode:2011ApJ...730...74M. arXiv:1007.3787. doi:10.1088/0004-637X/730/2/74

- Ghose, Tia (26 February 2015). "Big Bang, Deflated? Universe May Have Had No Beginning". Live Science. Retrieved 28 February 2015

- Friedmann, Alexander (1922), "Über die Krümmung des Raumes", Zeitschrift für Physik A, 10 (1): 377–386, Bibcode:1922ZPhy...10..377F, doi:10.1007/BF01332580

- See M. Sami; R. Myrzakulov (2015). "Late time cosmic acceleration: ABCD of dark energy and modified theories of gravity". International Journal of Modern Physics D. 25 (12). arXiv:1309.4188. doi:10.1142/S0218271816300317

- North J D:(1965)The Measure of the Universe - a history of modern cosmology, Oxford Univ. Press, Dover reprint 1990, ISBN 0-486-66517-8

- Ali, Ahmed Faraq (4 February 2015). "Cosmology from quantum potential". Physics Letters B. 741: 276–279. doi:10.1016/j.physletb.2014.12.057. Retrieved 28 February 2015

- Heilbron, J.L.: The Sun in the Church: Cathedrals as Solar Observatories. Cambridge, Massachusetts, Harvard University Press, 1999 ISBN 0-674-85433-0

- Overbye, Dennis (June 15, 2016). "Scientists Hear a Second Chirp From Colliding Black Holes". The New York Times. Retrieved June 15, 2016

- Peebles, P. J. E.; Ratra, Bharat (2003). "The cosmological constant and dark energy". Reviews of Modern Physics. 75 (2): 559–606. Bibcode:2003RvMP...75..559P. arXiv:astro-ph/0207347. doi:10.1103/RevMod-Phys.75.559

- Perlmutter, S.; et al. (June 1999). "Measurements of Omega and Lambda from 42 High-Redshift Supernovae". The Astrophysical Journal. 517 (2): 565–586. Bibcode:1999ApJ...517..565P. arXiv:astro-ph/9812133. doi:10.1086/307221

- MP Hobson; GP Efstathiou; AN Lasenby (2006). General Relativity: An introduction for physicists (Reprinted with corrections 2007 ed.). Cambridge University Press. p. 187. ISBN 978-0-521-82951-9

- "The Newest Search for Gravitational Waves has Begun". LIGO Caltech. LIGO. 18 September 2015. Retrieved 29 November 2015

- Zhong-Yue Wang (2016). "Modern Theory for Electromagnetic Metamaterials". Plasmonics. 11 (2): 503–508. doi:10.1007/s11468-015-0071-7

- Weinberg, Steven (1993). Dreams of a Final Theory: the search for the fundamental laws of nature. Vintage Press. p. 182. ISBN 0-09-922391-0

- Weiss, Achim. "Equilibrium and change: The physics behind Big Bang Nucleosynthesis". Einstein Online. Archived from the original on 8 February 2007. Retrieved 2007-02-24

- Exirifard, Q. (2010). "Phenomenological covariant approach to gravity". General Relativity and Gravitation. 43: 93–106. Bibcode:2011GReGr..43...93E. arXiv:0808.1962. doi:10.1007/s10714-010-1073-6

- Overbye, Dennis (20 February 2017). "Cosmos Controversy: The Universe Is Expanding, but How Fast?". New York Times. Retrieved 21 February 2017

Observable Universe: An Integrated Study

The observable universe consists of matter that can be observed from Earth. Particle horizon, Hubble volume, redshift survey and comoving distance are some of the significant and important topics related to cosmology. The following chapter unfolds its crucial aspects in a critical yet systematic manner.

Observable Universe

The observable universe is a spherical region of the Universe comprising all matter that can be observed from Earth at the present time, because light and other signals from these objects have had time to reach Earth since the beginning of the cosmological expansion. There are at least two trillion galaxies in the observable universe, containing more stars than all the grains of sand on planet Earth. Assuming the universe is isotropic, the distance to the edge of the observable universe is roughly the same in every direction. That is, the observable universe is a spherical volume (a ball) centered on the observer. Every location in the Universe has its own observable universe, which may or may not overlap with the one centered on Earth.

The word *observable* used in this sense does not depend on whether modern technology actually permits detection of radiation from an object in this region (or indeed on whether there is any radiation to detect). It simply indicates that it is possible *in principle* for light or other signals from the object to reach an observer on Earth. In practice, we can see light only from as far back as the time of photon decoupling in the recombination epoch. That is when particles were first able to emit photons that were not quickly re-absorbed by other particles. Before then, the Universe was filled with a plasma that was opaque to photons. The detection of gravitational waves indicates there is now a possibility of detecting non-light signals from before the recombination epoch.

The surface of last scattering is the collection of points in space at the exact distance that photons from the time of photon decoupling just reach us today. These are the photons we detect today as cosmic microwave background radiation (CMBR). However, with future technology, it may be possible to observe the still older relic neutrino background, or even more distant events via gravitational waves (which also should move at the speed of light). Sometimes astrophysicists distinguish between the *visible* universe, which includes only signals emitted since recombination – and the *observable* universe, which includes signals since the beginning of the cosmological expansion (the Big Bang in traditional physical cosmology, the end of the inflationary epoch in modern cosmology). According to calculations, the *comoving distance* (current proper distance) to particles from which the CMBR was emitted, which represent the radius of the visible universe, is about 14.0 billion parsecs (about 45.7 billion light years), while the comoving distance to the edge of the observable universe is about 14.3 billion parsecs (about 46.6 billion light years), about 2% larger.

The best estimate (as of 2015) of the age of the universe is 13.799±0.021 billion years but due to the expansion of space humans are observing objects that were originally much closer but are now considerably farther away (as defined in terms of cosmological proper distance, which is equal to the comoving distance at the present time) than a static 13.8 billion light-years distance. It is estimated that the diameter of the observable universe is about 28.5 gigaparsecs (93 billion light-years, 8.8×10^{23} kilometres or 5.5×10^{23} miles), putting the edge of the observable universe at about 46.5 billion light-years away. The total mass of ordinary matter in the universe can be calculated using the critical density and the diameter of the observable universe to be about 1.5×10^{53} kg.

The Universe Versus the Observable Universe

Some parts of the Universe are too far away for the light emitted since the Big Bang to have had enough time to reach Earth, so these portions of the Universe lie outside the observable universe. In the future, light from distant galaxies will have had more time to travel, so additional regions will become observable. However, due to Hubble's law, regions sufficiently distant from the Earth are expanding away from it faster than the speed of light (special relativity prevents nearby objects in the same local region from moving faster than the speed of light with respect to each other, but there is no such constraint for distant objects when the space between them is expanding;) and furthermore the expansion rate appears to be accelerating due to dark energy. Assuming dark energy remains constant (an unchanging cosmological constant), so that the expansion rate of the Universe continues to accelerate, there is a "future visibility limit" beyond which objects will *never* enter our observable universe at any time in the infinite future, because light emitted by objects outside that limit would never reach the Earth. (A subtlety is that, because the Hubble parameter is decreasing with time, there can be cases where a galaxy that is receding from the Earth just a bit faster than light does emit a signal that reaches the Earth eventually). This future visibility limit is calculated at a comoving distance of 19 billion parsecs (62 billion light years), assuming the Universe will keep expanding forever, which implies the number of galaxies that we can ever theoretically observe in the infinite future (leaving aside the issue that some may be impossible to observe in practice due to redshift, as discussed in the following paragraph) is only larger than the number currently observable by a factor of 2.36.

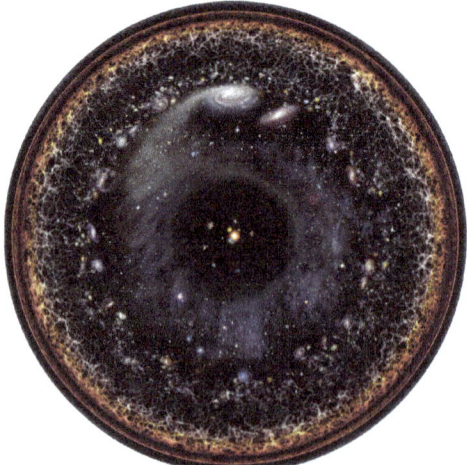

Artist's logarithmic scale conception of the *observable universe* with the Solar System at the center, inner and outer planets, Kuiper belt, Oort cloud, Alpha Centauri, Perseus Arm, Milky Way galaxy, Andromeda galaxy, nearby galaxies, Cosmic Web, Cosmic microwave radiation and the Big Bang's invisible plasma on the edge.

Though in principle more galaxies will become observable in the future, in practice an increasing number of galaxies will become extremely redshifted due to ongoing expansion, so much so that they will seem to disappear from view and become invisible. An additional subtlety is that a galaxy at a given comoving distance is defined to lie within the "observable universe" if we can receive signals emitted by the galaxy at any age in its past history (say, a signal sent from the galaxy only 500 million years after the Big Bang), but because of the Universe's expansion, there may be some later age at which a signal sent from the same galaxy can *never* reach the Earth at any point in the infinite future (so for example we might never see what the galaxy looked like 10 billion years after the Big Bang), even though it remains at the same comoving distance (comoving distance is defined to be constant with time – unlike proper distance, which is used to define recession velocity due to the expansion of space), which is less than the comoving radius of the observable universe. This fact can be used to define a type of cosmic event horizon whose distance from the Earth changes over time. For example, the current distance to this horizon is about 16 billion light years, meaning that a signal from an event happening *at present* can eventually reach the Earth in the future if the event is less than 16 billion light years away, but the signal will never reach the Earth if the event is more than 16 billion light years away.

Both popular and professional research articles in cosmology often use the term "universe" to mean "observable universe". This can be justified on the grounds that we can never know anything by direct experimentation about any part of the Universe that is causally disconnected from the Earth, although many credible theories require a total universe much larger than the observable universe. No evidence exists to suggest that the boundary of the observable universe constitutes a boundary on the Universe as a whole, nor do any of the mainstream cosmological models propose that the Universe has any physical boundary in the first place, though some models propose it could be finite but unbounded, like a higher-dimensional analogue of the 2D surface of a sphere that is finite in area but has no edge. It is plausible that the galaxies within our observable universe represent only a minuscule fraction of the galaxies in the Universe. According to the theory of cosmic inflation initially introduced by its founder, Alan Guth (and by D. Kazanas), if it is assumed that inflation began about 10^{-37} seconds after the Big Bang, then with the plausible assumption that the size of the Universe before the inflation occurred was approximately equal to the speed of light times its age, that would suggest that at present the entire universe's size is at least 3×10^{23} times the radius of the observable universe. There are also lower estimates claiming that the entire universe is in excess of 250 times larger than the observable universe and also higher estimates implying that the universe is at least $10^{10^{10^{122}}}$ times larger than the observable universe.

If the Universe is finite but unbounded, it is also possible that the Universe is *smaller* than the observable universe. In this case, what we take to be very distant galaxies may actually be duplicate images of nearby galaxies, formed by light that has circumnavigated the Universe. It is difficult to test this hypothesis experimentally because different images of a galaxy would show different eras in its history, and consequently might appear quite different. Bielewicz et al. claims to establish a lower bound of 27.9 gigaparsecs (91 billion light-years) on the diameter of the last scattering surface (since this is only a lower bound, the paper leaves open the possibility that the whole universe is much larger, even infinite). This value is based on matching-circle analysis of the WMAP 7 year data. This approach has been disputed.

Size

Hubble Ultra-Deep Field image of a region of the observable universe (equivalent sky area size shown in bottom left corner), near the constellation Fornax. Each spot is a galaxy, consisting of billions of stars. The light from the smallest, most red-shifted galaxies originated nearly 14 billion years ago.

The comoving distance from Earth to the edge of the observable universe is about 14.26 gigaparsecs (46.5 billion light years or 4.40×10^{26} meters) in any direction. The observable universe is thus a sphere with a diameter of about 28.5 gigaparsecs (93 Gly or 8.8×10^{26} m). Assuming that space is roughly flat, this size corresponds to a comoving volume of about 1.22×10^4 Gpc³ (4.22×10^5 Gly³ or 3.57×10^{80} m³).

The figures quoted above are distances *now* (in cosmological time), not distances *at the time the light was emitted*. For example, the cosmic microwave background radiation that we see right now was emitted at the time of photon decoupling, estimated to have occurred about 380000 years after the Big Bang, which occurred around 13.8 billion years ago. This radiation was emitted by matter that has, in the intervening time, mostly condensed into galaxies, and those galaxies are now calculated to be about 46 billion light-years from us. To estimate the distance to that matter at the time the light was emitted, we may first note that according to the Friedmann–Lemaître–Robertson–Walker metric, which is used to model the expanding universe, if at the present time we receive light with a redshift of z, then the scale factor at the time the light was originally emitted is given by

$$a(t) = \frac{1}{1+z}.$$

WMAP nine-year results combined with other measurements give the redshift of photon decoupling as $z = 1091.64 \pm 0.47$, which implies that the scale factor at the time of photon decoupling would be $\frac{1}{1092.64}$. So if the matter that originally emitted the oldest CMBR photons has a *present* distance of 46 billion light years, then at the time of decoupling when the photons were originally emitted, the distance would have been only about 42 *million* light-years.

Misconceptions on its Size

Many secondary sources have reported a wide variety of incorrect figures for the size of the visible universe. Some of these figures are listed below, with brief descriptions of possible reasons for misconceptions about them.

An example of one of the most common misconceptions about the size of the observable universe.
This plaque appears at the Rose Center for Earth and Space in New York City.

13.8 billion light-years

The age of the universe is estimated to be 13.8 billion years. While it is commonly understood that nothing can accelerate to velocities equal to or greater than that of light, it is a
common misconception that the radius of the observable universe must therefore amount
to only 13.8 billion light-years. This reasoning would only make sense if the flat, static
Minkowski spacetime conception under special relativity were correct. In the real universe,
spacetime is curved in a way that corresponds to the expansion of space, as evidenced by
Hubble's law. Distances obtained as the speed of light multiplied by a cosmological time
interval have no direct physical significance.

15.8 billion light-years

This is obtained in the same way as the 13.8 billion light year figure, but starting from an
incorrect age of the universe that the popular press reported in mid-2006. For an analysis
of this claim and the paper that prompted it.

27.6 billion light-years

This is a diameter obtained from the (incorrect) radius of 13.8 billion light-years.

78 billion light-years

In 2003, Cornish et al. found this lower bound for the diameter of the *whole* universe (not
just the observable part), if we postulate that the universe is finite in size due to its having a nontrivial topology, with this lower bound based on the estimated current distance
between points that we can see on opposite sides of the cosmic microwave background
radiation (CMBR). If the whole universe is smaller than this sphere, then light has had
time to circumnavigate it since the Big Bang, producing multiple images of distant points
in the CMBR, which would show up as patterns of repeating circles. Cornish et al. looked
for such an effect at scales of up to 24 gigaparsecs (78 Gly or 7.4×10^{26} m) and failed to find
it, and suggested that if they could extend their search to all possible orientations, they
would then «be able to exclude the possibility that we live in a universe smaller than 24
Gpc in diameter». The authors also estimated that with «lower noise and higher resolution
CMB maps (from WMAP's extended mission and from Planck), we will be able to search
for smaller circles and extend the limit to ~28 Gpc." This estimate of the maximum lower
bound that can be established by future observations corresponds to a radius of 14 giga-

parsecs, or around 46 billion light years, about the same as the figure for the radius of the visible universe (whose radius is defined by the CMBR sphere) given in the opening section. A 2012 preprint by most of the same authors as the Cornish et al. paper has extended the current lower bound to a diameter of 98.5% the diameter of the CMBR sphere, or about 26 Gpc.

156 billion light-years

This figure was obtained by doubling 78 billion light-years on the assumption that it is a radius. Because 78 billion light-years is already a diameter (the original paper by Cornish et al. says, "By extending the search to all possible orientations, we will be able to exclude the possibility that we live in a universe smaller than 24 Gpc in diameter," and 24 Gpc is 78 billion light years), the doubled figure is incorrect. This figure was very widely reported. A press release from Montana State University–Bozeman, where Cornish works as an astrophysicist, noted the error when discussing a story that had appeared in *Discover* magazine, saying "*Discover* mistakenly reported that the universe was 156 billion light-years wide, thinking that 78 billion was the radius of the universe instead of its diameter."

180 billion light-years

This estimate combines the erroneous 156-billion-light-year figure with evidence that the M33 Galaxy is actually fifteen percent farther away than previous estimates and that, therefore, the Hubble constant is fifteen percent smaller. The 180-billion figure is obtained by adding 15% to 156 billion light years.

Large-scale Structure

Sky surveys and mappings of the various wavelength bands of electromagnetic radiation (in particular 21-cm emission) have yielded much information on the content and character of the universe's structure. The organization of structure appears to follow as a hierarchical model with organization up to the scale of superclusters and filaments. Larger than this (at scales between 30-200 megaparsecs), there seems to be no continued structure, a phenomenon that has been referred to as the *End of Greatness*.

Walls, Filaments, Nodes, and Voids

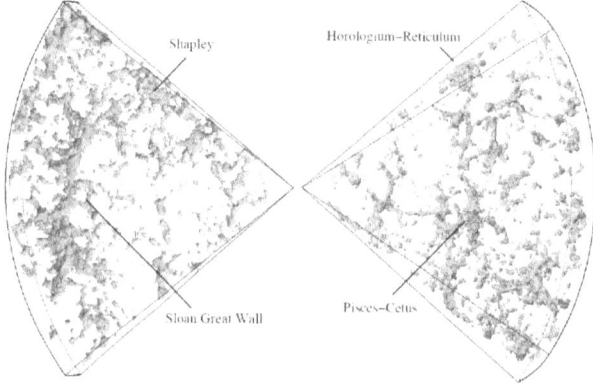

DTFE reconstruction of the inner parts of the 2dF Galaxy Redshift Survey

The organization of structure arguably begins at the stellar level, though most cosmologists rarely address astrophysics on that scale. Stars are organized into galaxies, which in turn form galaxy groups, galaxy clusters, superclusters, sheets, walls and filaments, which are separated by immense voids, creating a vast foam-like structure sometimes called the "cosmic web". Prior to 1989, it was commonly assumed that virialized galaxy clusters were the largest structures in existence, and that they were distributed more or less uniformly throughout the Universe in every direction. However, since the early 1980s, more and more structures have been discovered. In 1983, Adrian Webster identified the Webster LQG, a large quasar group consisting of 5 quasars. The discovery was the first identification of a large-scale structure, and has expanded the information about the known grouping of matter in the Universe. In 1987, Robert Brent Tully identified the Pisces–Cetus Supercluster Complex, the galaxy filament in which the Milky Way resides. It is about 1 billion light years across. That same year, an unusually large region with no galaxies was discovered, the Giant Void, which measures 1.3 billion light years across. Based on redshift survey data, in 1989 Margaret Geller and John Huchra discovered the "Great Wall", a sheet of galaxies more than 500 million light-years long and 200 million light-years wide, but only 15 million light-years thick. The existence of this structure escaped notice for so long because it requires locating the position of galaxies in three dimensions, which involves combining location information about the galaxies with distance information from redshifts. Two years later, astronomers Roger G. Clowes and Luis E. Campusano discovered the Clowes–Campusano LQG, a large quasar group measuring two billion light years at its widest point, and was the largest known structure in the Universe at the time of its announcement. In April 2003, another large-scale structure was discovered, the Sloan Great Wall. In August 2007, a possible supervoid was detected in the constellation Eridanus. It coincides with the 'CMB cold spot', a cold region in the microwave sky that is highly improbable under the currently favored cosmological model. This supervoid could cause the cold spot, but to do so it would have to be improbably big, possibly a billion light-years across, almost as big as the Giant Void mentioned above.

Computer simulated image of an area of space more than 50 million light years across, presenting a possible large-scale distribution of light sources in the universe - precise relative contributions of galaxies and quasars are unclear.

Another large-scale structure is the Newfound Blob, a collection of galaxies and enormous gas bubbles that measures about 200 million light years across.

In 2011, a large quasar group was discovered, U1.11, measuring about 2.5 billion light years across. On January 11, 2013, another large quasar group, the Huge-LQG, was discovered, which was measured to be four billion light-years across, the largest known structure in the Universe that time. In November 2013, astronomers discovered the Hercules–Corona Borealis Great Wall, an even bigger structure twice as large as the former. It was defined by the mapping of gamma-ray bursts.

End of Greatness

The *End of Greatness* is an observational scale discovered at roughly 100 Mpc (roughly 300 million lightyears) where the lumpiness seen in the large-scale structure of the universe is homogenized and isotropized in accordance with the Cosmological Principle. At this scale, no pseudo-random fractalness is apparent. The superclusters and filaments seen in smaller surveys are randomized to the extent that the smooth distribution of the Universe is visually apparent. It was not until the redshift surveys of the 1990s were completed that this scale could accurately be observed.

Observations

"Panoramic view of the entire near-infrared sky reveals the distribution of galaxies beyond the Milky Way. The image is derived from the 2MASS Extended Source Catalog (XSC) – more than 1.5 million galaxies, and the Point Source Catalog (PSC) – nearly 0.5 billion Milky Way stars. The galaxies are color-coded by ‹redshift› obtained from the UGC, CfA, Tully NBGC, LCRS, 2dF, 6dFGS, and SDSS surveys (and from various observations compiled by the NASA Extragalactic Database), or photo-metrically deduced from the K band (2.2 μm). Blue are the nearest sources (z < 0.01); green are at moderate distances (0.01 < z < 0.04) and red are the most distant sources that 2MASS resolves (0.04 < z < 0.1). The map is projected with an equal area Aitoff in the Galactic system (Milky Way at center)."

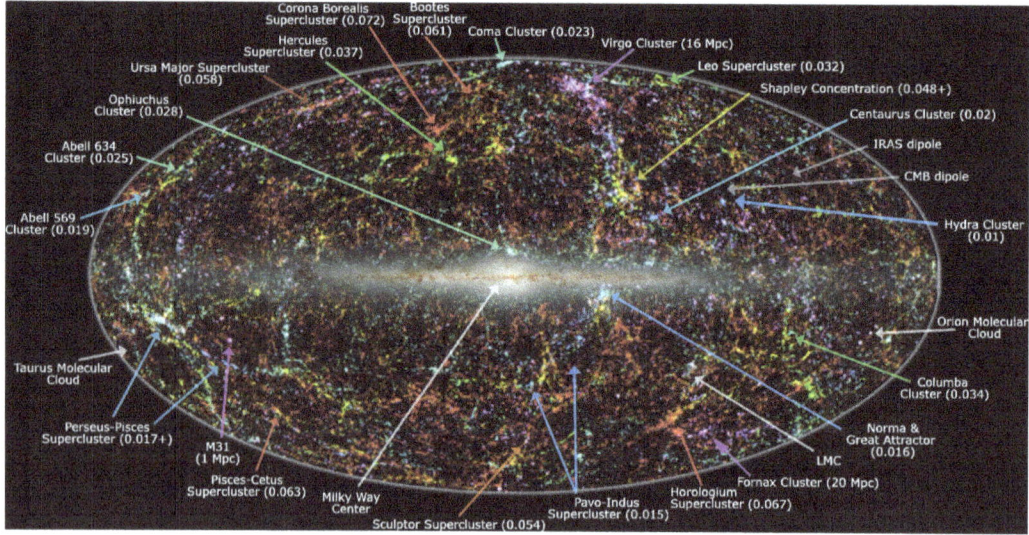

"Panoramic view of the entire near-infrared sky reveals the distribution of galaxies beyond the Milky Way. The image is derived from the 2MASS Extended Source Catalog (XSC) – more than 1.5 million galaxies, and the Point Source Catalog (PSC) – nearly 0.5 billion Milky Way stars. The galaxies are color-coded by ‘redshift’ obtained from the UGC, CfA, Tully NBGC, LCRS, 2dF, 6dFGS, and SDSS surveys (and from various observations compiled by the NASA Extragalactic Database), or photo-metrically deduced from the K band (2.2 μm). Blue are the nearest sources (z < 0.01); green are at moderate distances (0.01 < z < 0.04) and red are the most distant sources that 2MASS resolves (0.04 < z < 0.1). The map is projected with an equal area Aitoff in the Galactic system (Milky Way at center)."

Another indicator of large-scale structure is the 'Lyman-alpha forest'. This is a collection of absorption lines that appear in the spectra of light from quasars, which are interpreted as indicating the existence of huge thin sheets of intergalactic (mostly hydrogen) gas. These sheets appear to be associated with the formation of new galaxies.

Caution is required in describing structures on a cosmic scale because things are often different from how they appear. Gravitational lensing (bending of light by gravitation) can make an image appear to originate in a different direction from its real source. This is caused when foreground objects (such as galaxies) curve surrounding spacetime (as predicted by general relativity), and deflect passing light rays. Rather usefully, strong gravitational lensing can sometimes magnify distant galaxies, making them easier to detect. Weak lensing (gravitational shear) by the intervening universe in general also subtly changes the observed large-scale structure. As of 2004, measurements of this subtle shear showed considerable promise as a test of cosmological models.

The large-scale structure of the Universe also looks different if one only uses redshift to measure distances to galaxies. For example, galaxies behind a galaxy cluster are attracted to it, and so fall towards it, and so are slightly blueshifted (compared to how they would be if there were no cluster) On the near side, things are slightly redshifted. Thus, the environment of the cluster looks a bit squashed if using redshifts to measure distance. An opposite effect works on the galaxies already within a cluster: the galaxies have some random motion around the cluster center, and when these random motions are converted to redshifts, the cluster appears elongated. This creates a *"finger of God"* – the illusion of a long chain of galaxies pointed at the Earth.

Cosmography of Our Cosmic Neighborhood

At the centre of the Hydra-Centaurus Supercluster, a gravitational anomaly called the Great Attractor affects the motion of galaxies over a region hundreds of millions of light-years across. These galaxies are all redshifted, in accordance with Hubble's law. This indicates that they are receding from us and from each other, but the variations in their redshift are sufficient to reveal the existence of a concentration of mass equivalent to tens of thousands of galaxies.

The Great Attractor, discovered in 1986, lies at a distance of between 150 million and 250 million light-years (250 million is the most recent estimate), in the direction of the Hydra and Centaurus constellations. In its vicinity there is a preponderance of large old galaxies, many of which are colliding with their neighbours, or radiating large amounts of radio waves.

In 1987, astronomer R. Brent Tully of the University of Hawaii's Institute of Astronomy identified what he called the Pisces–Cetus Supercluster Complex, a structure one billion light years long and 150 million light years across in which, he claimed, the Local Supercluster was embedded.

Mass of Ordinary Matter

The mass of the observable Universe is often quoted as 10^{50} tonnes or 10^{53} kg. In this context, mass refers to ordinary matter and includes the interstellar medium (ISM) and the intergalactic medium (IGM). However, it excludes dark matter and dark energy. This quoted value for the mass of ordinary matter in the Universe can be estimated based on critical density. The calculations are for the observable universe only as the volume of the whole is unknown and may be infinite.

Estimates Based on Critical Density

Critical Density is the energy density where the expansion of the Universe is poised between continued expansion and collapse. Observations of the cosmic microwave background from the

Wilkinson Microwave Anisotropy Probe suggest that the spatial curvature of the Universe is very close to zero, which in current cosmological models implies that the value of the density parameter must be very close to a certain critical density value. At this condition, the calculation for ρ_c critical density, is:

$$\rho_c = \frac{3H_0^2}{8\pi G}$$

where G is the gravitational constant. From The European Space Agency's Planck Telescope results: H_0, is 67.15 kilometers per second per mega parsec. This gives a critical density of 0.85×10^{-26} kg/ m^3 (commonly quoted as about 5 hydrogen atoms per cubic meter). This density includes four significant types of energy/mass: ordinary matter (4.8%), neutrinos (0.1%), cold dark matter (26.8%), and dark energy (68.3%). Note that although neutrinos are defined as particles like electrons, they are listed separately because they are difficult to detect and so different from ordinary matter. Thus, the density of ordinary matter is 4.8% of the total critical density calculated or 4.08×10^{-28} kg/m^3. To convert this density to mass we must multiply by volume, a value based on the radius of the "observable universe". Since the Universe has been expanding for 13.8 billion years, the comoving distance (radius) is now about 46.6 billion light years. Thus, volume $(4/3\pi r^3)$ equals 3.58×10^{80} m^3 and mass of ordinary matter equals density $(4.08 \times 10^{-28}$ kg/m$^3)$ times volume $(3.58 \times 10^{80}$ m$^3)$ or 1.46×10^{53} kg.

Matter Content – Number of Atoms

Assuming the mass of ordinary matter is about 1.45×10^{53} kg and assuming all atoms are hydrogen atoms (which in reality make up about 74% of all atoms in our galaxy by mass), calculating the estimated total number of atoms in the observable Universe is straightforward. Divide the mass of ordinary matter by the mass of a hydrogen atom (1.45×10^{53} kg divided by 1.67×10^{-27} kg). The result is approximately 10^{80} hydrogen atoms. The chemistry of life may have begun shortly after the Big Bang, 13.8 billion years ago, during a habitable epoch when the Universe was only 10–17 million years old. According to the panspermia hypothesis, microscopic life – distributed by meteoroids, asteroids and other small Solar System bodies – may exist throughout the Universe. Though life is confirmed only on the Earth, many think that extraterrestrial life is not only plausible, but probable or inevitable.

Most Distant Objects

The most distant astronomical object yet announced as of January 2011 is a galaxy candidate classified UDFj-39546284. In 2009, a gamma ray burst, GRB 090423, was found to have a redshift of 8.2, which indicates that the collapsing star that caused it exploded when the Universe was only 630 million years old. The burst happened approximately 13 billion years ago, so a distance of about 13 billion light years was widely quoted in the media (or sometimes a more precise figure of 13.035 billion light years), though this would be the "light travel distance" rather than the "proper distance" used in both Hubble's law and in defining the size of the observable universe (cosmologist Ned Wright argues against the common use of light travel distance in astronomical press releases on this page, and at the bottom of the page offers online calculators that can be used to calculate the current proper distance to a distant object in a flat universe based on either the redshift z

or the light travel time). The proper distance for a redshift of 8.2 would be about 9.2 Gpc, or about 30 billion light years. Another record-holder for most distant object is a galaxy observed through and located beyond Abell 2218, also with a light travel distance of approximately 13 billion light years from Earth, with observations from the Hubble telescope indicating a redshift between 6.6 and 7.1, and observations from Keck telescopes indicating a redshift towards the upper end of this range, around 7. The galaxy's light now observable on Earth would have begun to emanate from its source about 750 million years after the Big Bang.

Horizons

The limit of observability in our universe is set by a set of cosmological horizons which limit—based on various physical constraints—the extent to which we can obtain information about various events in the Universe. The most famous horizon is the particle horizon which sets a limit on the precise distance that can be seen due to the finite age of the Universe. Additional horizons are associated with the possible future extent of observations (larger than the particle horizon owing to the expansion of space), an "optical horizon" at the surface of last scattering, and associated horizons with the surface of last scattering for neutrinos and gravitational waves.

A diagram of our location in the observable universe

Particle Horizon

The particle horizon (also called the cosmological horizon, the comoving horizon (in Dodelson's text), or the cosmic light horizon) is the maximum distance from which particles could have traveled to the observer in the age of the universe. Much like the concept of a terrestrial horizon, it represents the boundary between the observable and the unobservable regions of the universe, so its distance at the present epoch defines the size of the observable universe. Due to the expansion of the universe it is not simply the age of the universe times the speed of light (approximately 13.8 billion years), but rather the speed of light times the conformal time. The existence, properties, and significance of a cosmological horizon depend on the particular cosmological model.

Conformal Time and the Particle Horizon

In terms of comoving distance, the particle horizon is equal to the conformal time η that has passed since the Big Bang, times the speed of light c. In general, the conformal time at a certain time t is given by

$$\eta = \int_0^t \frac{dt'}{a(t')},$$

where $a(t)$ is the scale factor of the Friedmann–Lemaître–Robertson–Walker metric, and we have taken the Big Bang to be at $t = 0$. By convention, a subscript 0 indicates "today" so that the conformal time today $\eta(t_0) = \eta_0 = 1.48 \times 10^{18}$ s. Note that the conformal time is not the age of the universe. Rather, the conformal time is the amount of time it would take a photon to travel from where we are located to the furthest observable distance provided the universe ceased expanding. As such, η_0 is not a physically meaningful time (this much time has not yet actually passed), though, as we will see, the particle horizon with which it is associated is a conceptually meaningful distance.

The particle horizon recedes constantly as time passes and the conformal time grows. As such, the observed size of the universe always increases. Since proper distance at a given time is just comoving distance times the scale factor (with comoving distance normally defined to be equal to proper distance at the present time, so $a(t_0) = 1$ at present), the proper distance to the particle horizon at time t is given by

$$a(t)H_p(t) = a(t)\int_0^t \frac{cdt'}{a(t')}$$

and for today $t = t_0$

$$. \ H_p(t_0) = c\eta_0 = 14.4 \text{ Gpc} = 46.9 \text{ billion light years.}$$

Evolution of the Particle Horizon

We consider the FLRW cosmological model. In that context, the universe can be approximated as composed by non-interacting constituents, each one being a perfect fluid with density ρ_i, partial pressure p_i and state equation $p_i = \omega_i \rho_i$, such that they add up to the total density ρ and total pressure p. Let us now define the following functions:

- Hubble function $H = \dfrac{\dot{a}}{a}$

- The critical density $\rho_c = \dfrac{3}{8\pi} H^2$

- The i-th dimensionless energy density $\Omega_i = \dfrac{\rho_i}{\rho_c}$

- The dimensionless energy density $\Omega = \dfrac{\rho}{\rho_c} = \sum \Omega_i$

- The redshift z given by the formula $1 + z = \dfrac{a_0}{a(t)}$

Any function with a zero subscript denote the function evaluated at the present time t_0 (or equivalently $z = 0$). The last term can be taken to be 1 including the curvature state equation. It can be proved that the Hubble function is given by

$$H(z) = H_0 \sqrt{\sum \Omega_{i0}(1+z)^{n_i}}$$

where $n_i = 3(1 + \omega_i)$. Notice that the addition ranges over all possible partial constituents and in particular there can be countably infinitely many. With this notation we have:

The particle horizon H_p exists if and only if $N > 2$

where N is the largest n_i (possibly infinite). The evolution of the particle horizon for an expanding universe ($\dot{a} > 0$) is:

$$\frac{dH_p}{dt} = H_p(z)H(z) + c$$

where c is the speed of light and can be taken to be 1 (natural units). Notice that the derivative is made with respect to the FLRW-time t, while the functions are evaluated at the redshift z which are related as stated before. We have an analogous but slightly different result for Event Horizon.

Horizon Problem

The concept of a particle horizon can be used to illustrate the famous horizon problem, which is an unresolved issue associated with the Big Bang model. Extrapolating back to the time of recombination when the cosmic microwave background (CMB) was emitted, we obtain a particle horizon of about

$$H_p(t_{\mathrm{CMB}}) = c\eta_{\mathrm{CMB}} = 284 \text{ Mpc} = 8.9 \times 10^{-3} H_p(t_0).$$

which corresponds to a proper size at that time of:

$$a_{\mathrm{CMB}} H_p(t_{\mathrm{CMB}}) = 261 \text{ kpc}$$

Since we observe the CMB to be emitted essentially from our particle horizon (284 Mpc $\ll 14.4$ Gpc), our expectation is that parts of the cosmic microwave background (CMB) that are separated by about a fraction of a great circle across the sky of

$$f = H_p(t_{\mathrm{CMB}}) / H_p(t_0)$$

(an angular size of $\theta \sim 1.7°$) should be out of causal contact with each other. That the entire CMB is in thermal equilibrium and approximates a blackbody so well is therefore not explained by the standard explanations about the way the expansion of the universe proceeds. The most popular resolution to this problem is cosmic inflation.

Hubble Volume

In cosmology, a Hubble volume, or Hubble sphere, is a spherical region of the Universe surrounding an observer beyond which objects recede from that observer at a rate greater than the speed of light due to the expansion of the Universe. The Hubble volume is approximately equal to 10^{31} cubic light years.

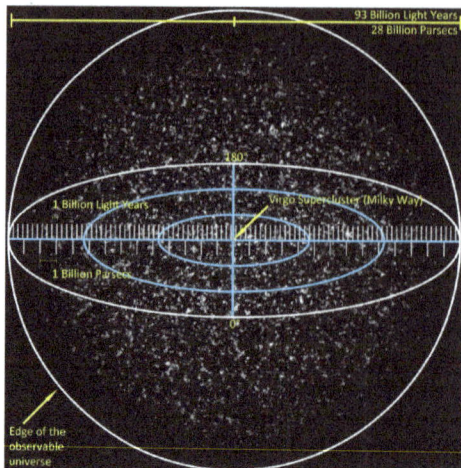

Visualization of the three-dimensional large-scale structure of the universe in the Hubble volume. The scale is such that the fine grains of light represent collections of large numbers of superclusters. The Virgo Supercluster - home of our own galaxy - is at the center of our Hubble volume, but is too small to be seen in the image.

The proper radius of a Hubble sphere (known as the Hubble radius or the Hubble length) is c/H_0, where c is the speed of light and H_0 is the Hubble constant. The surface of a Hubble sphere is called the *microphysical horizon*, the *Hubble surface*, or the *Hubble limit*.

More generally, the term "Hubble volume" can be applied to any region of space with a volume of order $(c/H_0)^3$. However, the term is also frequently (but mistakenly) used as a synonym for the observable universe; the latter is larger than the Hubble volume.

Relationship to age of the Universe

The Hubble length c/H_0 is 14 billion light years in the standard cosmological model, somewhat larger than c times the age of the universe, 13.8 billion years.

Hubble Limit as an Event Horizon

Objects at the Hubble limit have an average proper speed of c relative to an observer on the Earth so that, in a universe with constant Hubble parameter, light emitted at the present time by objects outside the Hubble limit would never be seen by an observer on Earth. That is, the Hubble limit would coincide with a cosmological event horizon (a boundary separating events visible at some time and those that are never visible).

However, the Hubble parameter is not constant in various cosmological models so that the Hubble limit does not, in general, coincide with a cosmological event horizon. For example in a decelerating Friedmann universe the Hubble sphere expands faster than the Universe and its boundary overtakes light emitted by receding galaxies so that light emitted at earlier times by objects outside the Hubble sphere still may eventually arrive inside the sphere and be seen by us. Conversely, in an accelerating universe, the Hubble sphere expands more slowly than the Universe, and bodies move out of the Hubble sphere.

Observations indicate that the universe is accelerating, so that some objects that we can currently exchange signals with will one day cross our Hubble limit.

List of Cosmological Horizons

A cosmological horizon is a measure of the distance from which one could possibly retrieve information. This observable constraint is due to various properties of general relativity, the expanding universe, and the physics of Big Bang cosmology. Cosmological horizons set the size and scale of the observable universe.

Particle Horizon

The particle horizon (also called the cosmological horizon, the comoving horizon, or the cosmic light horizon) is the maximum distance from which particles could have traveled to the observer in the age of the universe. It represents the boundary between the observable and the unobservable regions of the universe, so its distance at the present epoch defines the size of the observable universe. Due to the expansion of the universe it is not simply the age of the universe times the speed of light, as in the Hubble horizon, but rather the speed of light multiplied by the conformal time. The existence, properties, and significance of a cosmological horizon depend on the particular cosmological model.

In terms of comoving distance, the particle horizon is equal to the conformal time that has passed since the Big Bang, times the speed of light. In general, the conformal time at a certain time is given in terms of the scale factor a by,

$$\eta(t) = \int_0^t \frac{dt'}{a(t')}$$

or

$$\eta(a) = \int_0^a \frac{1}{a'H(a')} \frac{da'}{a'}.$$

The particle horizon is the boundary between two regions at a point at a given time: one region defined by events that have already been observed by an observer, and the other by events which cannot be observed *at that time*. It represents the furthest distance from which we can retrieve information from the past, and so defines the observable universe.

Hubble Horizon

Hubble radius, Hubble sphere, Hubble volume, or Hubble horizon is a conceptual horizon defining the boundary between particles that are moving slower and faster than the speed of light relative to an observer at one given time. Note that this does not mean the particle is unobservable, the light from the past is reaching and will continue to reach the observer for a while.

The Hubble velocity of an object is given by Hubble's law,

$$v = xH.$$

Replacing v with speed of light c and solving for proper distance x we obtain the radius of Hubble sphere as

$$r_{HS}(t) = \frac{c}{H(t)}.$$

In an ever-accelerating universe, if two particles are separated by a distance greater than the Hubble radius, they cannot talk to each other from now on (as they are now, not as they have been in the past), However, if they are outside of each other's particle horizon, they could have never communicated. Depending on the form of expansion of the universe, they may be able to exchange information in the future. Today,

$$r_{HS}(t_0) = \frac{c}{H_0},$$

yielding a Hubble horizon of some 4.1 Gpc. This horizon is not really a physical size, but it is often used as useful length scale as most physical sizes in cosmology can be written in terms of those factors.

One can also define comoving Hubble horizon by simply dividing Hubble radius by the scale factor

$$r_{HS,comoving}(t) = \frac{c}{a(t)H(t)}.$$

Event Horizon

The particle horizon differs from the cosmic event horizon, in that the particle horizon represents the largest comoving distance from which light could have reached the observer by a specific time, while the event horizon is the largest comoving distance from which light emitted now can *ever* reach the observer in the future. The current distance to our cosmic event horizon is about 5 Gpc, well within our observable range given by the particle horizon.

In general, the proper distance to the event horizon at time t is given by

$$d_e(t) = a(t)\int_t^{t_{max}} \frac{cdt'}{a(t')}$$

where t_{max} is the time-coordinate of the end of the universe, which would be infinite in the case of a universe that expands forever.

For our case, assuming that dark energy is due to a cosmological constant, $d_e(t_0) < \infty$.

Future Horizon

In an accelerating universe, there are events which will be unobservable as $t \to \infty$ as signals from future events become redshifted to arbitrarily long wavelengths in the exponentially expanding de Sitter space. This sets a limit on the farthest distance that we can possibly see as measured in units of proper distance today. Or, more precisely, there are events that are spatially separated for a certain frame of reference happening simultaneously with the event occurring right now for which no signal will ever reach us, even though we can observe events that occurred at the same location in space that happened in the distant past. While we will continue to receive signals from this location in space, even if we wait an infinite amount of time, a signal that left from that location today will never reach us. Additionally, the signals coming from that location will have less and less energy and be less and less frequent until the location, for all practical purposes, becomes unobservable. In a universe that is dominated by dark energy which is undergoing an exponential expansion of the scale factor, all objects that are gravitationally unbound with respect to the Milky Way will become unobservable, in a futuristic version of Kapteyn's universe.

Practical Horizons

While not technically "horizons" in the sense of an impossibility for observations due to relativity or cosmological solutions, there are practical horizons which include the optical horizon, set at the surface of last scattering. This is the farthest distance that any photon can freely stream. Similarly, there is a "neutrino horizon" set for the farthest distance a neutrino can freely stream and a gravitational wave horizon at the farthest distance that gravitational waves can freely stream. The latter is predicted to be a direct probe of the end of cosmic inflation.

Redshift Survey

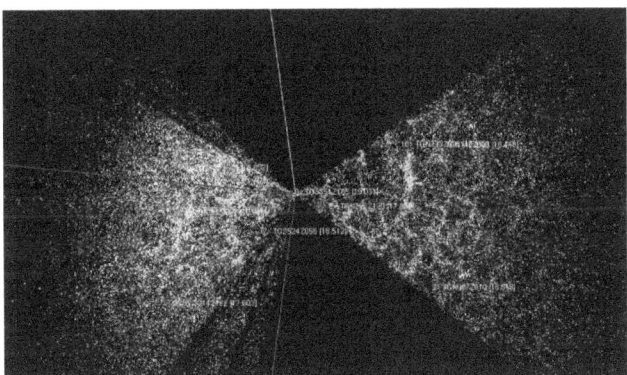

Rendering of the 2dFGRS data.

In astronomy, a redshift survey is a survey of a section of the sky to measure the redshift of astronomical objects: usually galaxies, but sometimes other objects such as galaxy clusters or quasars. Using Hubble's law, the redshift can be used to estimate the distance of an object from Earth. By combining redshift with angular position data, a redshift survey maps the 3D distribution of matter within a field of the sky. These observations are used to measure detailed statistical properties of the large-scale structure of the universe. In conjunction with observations of early structure in the cosmic microwave background, these results can place strong constraints on cosmological parameters such as the average matter density and the Hubble constant.

Generally the construction of a redshift survey involves two phases: first the selected area of the sky is imaged with a wide-field telescope, then galaxies brighter than a defined limit are selected from the resulting images as non-pointlike objects; optionally, colour selection may also be used to assist discrimination between stars and galaxies. Secondly, the selected galaxies are observed by spectroscopy, most commonly at visible wavelengths, to measure the wavelengths of prominent spectral lines; comparing observed and laboratory wavelengths then gives the redshift for each galaxy.

The Great Wall, a vast conglomeration of galaxies over 500 million light-years wide, provides a dramatic example of a large-scale structure that redshift surveys can detect.

The first systematic redshift survey was the CfA Redshift Survey of around 2,200 galaxies, started in 1977 with the initial data collection completed in 1982. This was later extended to the CfA2 redshift survey of 15,000 galaxies, completed in the early 1990s.

These early redshift surveys were limited in size by taking a spectrum for one galaxy at a time; from the 1990s, the development of fibre-optic spectrographs and multi-slit spectrographs enabled spectra for several hundred galaxies to be observed simultaneously, and much larger redshift surveys became feasible. Notable examples are the 2dF Galaxy Redshift Survey (221,000 redshifts, completed 2002); the Sloan Digital Sky Survey (approximately 1 million redshifts by 2007) and the Galaxy And Mass Assembly survey. At high redshift the largest current surveys are the DEEP2 Redshift Survey and the VIMOS-VLT Deep Survey (VVDS); these have around 50,000 redshifts each, and are mainly focused on galaxy evolution.

ZFOURGE or the FourStar Galaxy Evolution Survey is a large and deep medium-band imaging survey which aims to establish an observational benchmark of galaxy properties at redshift z > 1. The survey is using near-infrared FOURSTAR instrument on the Magellan Telescopes, surveying in all three HST legacy fields: COSMOS, CDFS, and UDS.

Because of the demands on observing time required to obtain spectroscopic redshifts (i.e., redshifts determined directly from spectral features measured at high precision), a common alternative is to use photometric redshifts based on model fits to the brightnesses and colors of objects. Such "photo-z's" can be used in large surveys to estimate the spatial distribution of galaxies and quasars, provided the galaxy types and colors are well understood in a particular redshift range. At present, the errors on photometric redshift measurements are significantly higher than those of spectroscopic redshifts, but future surveys (for example, the LSST) aim to significantly refine the technique.

Comoving Distance

In standard cosmology, comoving distance and proper distance are two closely related distance measures used by cosmologists to define distances between objects. *Proper distance* roughly corresponds to where a distant object would be at a specific moment of cosmological time, which can change over time due to the expansion of the universe. *Comoving distance* factors out the expansion of the universe, giving a distance that does not change in time due to the expansion of space (though this may change due to other, local factors, such as the motion of a galaxy within a cluster). Comoving distance and proper distance are defined to be equal at the present time; therefore, the ratio of proper distance to comoving distance now is 1. At other times, the scale factor differs from 1. The Universe's expansion results in the proper distance changing, while the comoving distance is unchanged by this expansion because it is the proper distance divided by that scale factor.

Comoving Coordinates

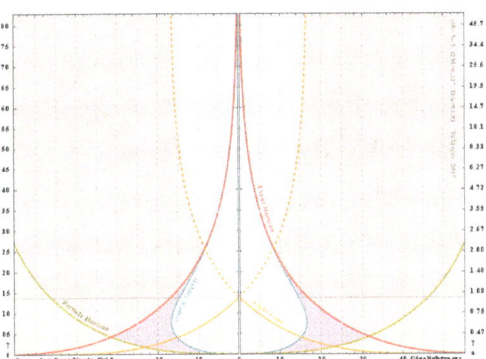

The evolution of the universe and its horizons in comoving distances

Although general relativity allows one to formulate the laws of physics using arbitrary coordinates, some coordinate choices are more natural or easier to work with. Comoving coordinates are an example of such a natural coordinate choice. They assign constant spatial coordinate values to observers who perceive the universe as isotropic. Such observers are called "comoving" observers because they move along with the Hubble flow.

A comoving observer is the only observer that will perceive the universe, including the cosmic microwave background radiation, to be isotropic. Non-comoving observers will see regions of the sky systematically blue-shifted or red-shifted. Thus isotropy, particularly isotropy of the cosmic microwave background radiation, defines a special local frame of reference called the comoving frame. The velocity of an observer relative to the local comoving frame is called the peculiar velocity of the observer.

Most large lumps of matter, such as galaxies, are nearly comoving, so that their peculiar velocities (owing to gravitational attraction) are low.

The comoving time coordinate is the elapsed time since the Big Bang according to a clock of a comoving observer and is a measure of cosmological time. The comoving spatial coordinates tell where an event occurs while cosmological time tells when an event occurs. Together, they form a complete coordinate system, giving both the location and time of an event.

Space in comoving coordinates is usually referred to as being "static", as most bodies on the scale of galaxies or larger are approximately comoving, and comoving bodies have static, unchanging comoving coordinates. So for a given pair of comoving galaxies, while the proper distance between them would have been smaller in the past and will become larger in the future due to the expansion of space, the comoving distance between them remains *constant* at all times.

The expanding Universe has an increasing scale factor which explains how constant comoving distances are reconciled with proper distances that increase with time.

Comoving Distance and Proper Distance

Comoving distance is the distance between two points measured along a path defined at the present cosmological time. For objects moving with the Hubble flow, it is deemed to remain constant in time. The comoving distance from an observer to a distant object (e.g. galaxy) can be computed by the following formula:

$$\chi = \int_{t_e}^{t} C\,\frac{dt'}{a(t')}$$

where $a(t')$ is the scale factor, t_e is the time of emission of the photons detected by the observer, t is the present time, and c is the speed of light in vacuum.

Despite being an integral over time, this does give the distance that *would* be measured by a hypothetical tape measure at *fixed* time t, i.e. the "proper distance" as defined below, divided by the scale factor $a(t)$ at that time. For a derivation see "standard relativistic definitions" from Davis & Lineweaver 2004.

Definitions

- Many textbooks use the symbol χ for the comoving distance. However, this χ must be distinguished from the *coordinate* distance r in the commonly used comoving coordinate system for a FLRW universe where the metric takes the form

$$ds^2 = -c^2 d\tau^2 = -c^2 dt^2 + a(t)^2 \left(\frac{dr^2}{1-kr^2} + r^2\left(d\theta^2 + \sin^2\theta\, d\phi^2 \right) \right).$$

  In this case the comoving coordinate distance r is related to χ by:

$$\chi = \begin{cases} \sinh^{-1} r, & \text{if } k=-1 \text{ (a negatively curved hyperbolic universe)} \\ r, & \text{if } k=0 \text{ (a spatially flat universe)} \\ \sin^{-1} r, & \text{if } k=1 \text{ (a positively curved spherical universe)} \end{cases}$$

- Most textbooks and research papers define the comoving distance between comoving observers to be a fixed unchanging quantity independent of time, while calling the dynamic, changing distance between them proper distance. On this usage, comoving and proper distances are numerically equal at the current age of the universe, but will differ in the past and in the future; if the comoving distance to a galaxy is denoted X, the proper distance $d(t)$ at an arbitrary time t is simply given by $d(t) = a(t)\chi$ where $a(t)$ is the scale factor (e.g. Davis & Limeweaver 2004). The proper distance $d(t)$ between two galaxies at time t is just the distance that would be measured by rulers between them at that time.

Uses of the Proper Distance

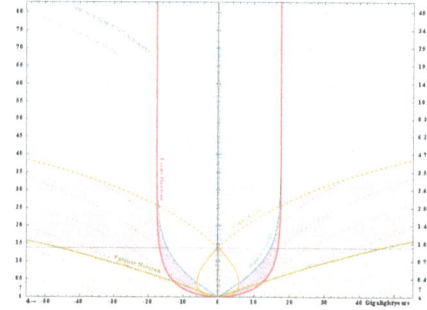

The evolution of the universe and its horizons in proper distances

Cosmological time is identical to locally measured time for an observer at a fixed comoving spatial position, that is, in the local comoving frame. Proper distance is also equal to the locally measured distance in the comoving frame for *nearby* objects. To measure the proper distance between two *distant* objects, one imagines that one has many comoving observers in a straight line between the two objects, so that all of the observers are close to each other, and form a chain between the two distant objects. All of these observers must have the same cosmological time. Each observer measures their distance to the nearest observer in the chain, and the length of the chain, the sum of distances between nearby observers, is the total proper distance.

It is important to the definition of both comoving distance and proper distance in the cosmological sense (as opposed to proper length in special relativity) that all observers have the same cosmological age. For instance, if one measured the distance along a straight line or spacelike geodesic between the two points, observers situated between the two points would have different cosmological ages when the geodesic path crossed their own world lines, so in calculating the distance along this geodesic one would not be correctly measuring comoving distance or cosmological proper distance. Comoving and proper distances are not the same concept of distance as the concept of distance in special relativity. This can be seen by considering the hypothetical case of a universe empty of mass, where both sorts of distance can be measured. When the density of mass in the FLRW metric is set to zero (an empty 'Milne universe'), then the cosmological coordinate system used to write this metric becomes a non-inertial coordinate system in the Minkowski spacetime of special relativity where surfaces of constant Minkowski proper-time τ appear as hyperbolas in the Minkowski diagram from the perspective of an inertial frame of reference. In this case, for two events which are simultaneous according to the cosmological time coordinate, the value of the cosmological proper distance is not equal to the value of the proper length between these same events, which would just be the distance along a straight line between the events in a Minkowski diagram (and a straight line is a geodesic in flat Minkowski spacetime), or the coordinate distance between the events in the inertial frame where they are simultaneous.

If one divides a change in proper distance by the interval of cosmological time where the change was measured (or takes the derivative of proper distance with respect to cosmological time) and calls this a "velocity", then the resulting "velocities" of galaxies or quasars can be above the speed of light, c. This apparent superluminal expansion is not in conflict with special or general relativity, and is a consequence of the particular definitions used in physical cosmology. Even light itself does not have a "velocity" of c in this sense; the total velocity of any object can be expressed as the sum $v_{tot} = v_{rec} + v_{pec}$ where v_{rec} is the recession velocity due to the expansion of the universe (the velocity given by Hubble's law) and v_{pec} is the "peculiar velocity" measured by local observers (with $v_{rec} = \dot{a}(t)\chi(t)$ and $v_{pec} = a(t)\dot{\chi}(t)$, the dots indicating a first derivative), so for light v_{pec} is equal to c ($-c$ if the light is emitted towards our position at the origin and $+c$ if emitted away from us) but the total velocity v_{tot} is generally different from c. Even in special relativity the coordinate speed of light is only guaranteed to be c in an inertial frame; in a non-inertial frame the coordinate speed may be different from c. In general relativity no coordinate system on a large region of curved spacetime is "inertial", but in the local neighborhood of any point in curved spacetime we can define a "local inertial frame" in which the local speed of light is c and in which massive objects such as stars and galaxies always have a local speed smaller than c. The cosmological definitions used to define the velocities of distant objects are coordinate-dependent - there is no general coordinate-independent definition of velocity between distant objects in general relativity. The issue of how best to describe and popularize the apparent superluminal expansion of the universe has caused a minor amount of controversy. One viewpoint is presented in Davis and Lineweaver, 2004.

Short Distances vs. Long Distances

Within small distances and short trips, the expansion of the universe during the trip can be ignored. This is because the travel time between any two points for a non-relativistic moving particle will just be the proper distance (that is, the comoving distance measured using the scale factor of the universe at the time of the trip rather than the scale factor "now") between those points divided by the velocity of the particle. If the particle is moving at a relativistic velocity, the usual relativistic corrections for time dilation must be made.

Chronology of the Universe

The chronology of the universe describes the history and future of the universe according to Big Bang cosmology. The metric expansion of space is estimated to have begun 13.8 billion years ago. For the purposes of this summary, it is convenient to divide the chronology of the universe into four parts:

1. The very early universe, from the Planck epoch until the cosmic inflation, or the first picosecond of cosmic time; this period is the domain of active theoretical research, currently beyond the grasp of most experiments in particle physics.

2. The early universe, from the Quark epoch to the Photon epoch, or the first 380,000 years of cosmic time, when the familiar forces and elementary particles have emerged but the universe remains in the state of a plasma, followed by the "Dark Ages", from 380,000 years to about 150 million years during which the universe was transparent but no large-scale structures had yet formed.

3. The period of large-scale structure formation, including stellar evolution, galaxy formation and evolution and the formation of galaxy clusters and superclusters, from about 150 million years to present, and prospectively until about 100 billion years of cosmic time; The thin disk of our galaxy began to form at about 5 billion years. The solar system formed at about 4.6 billion years ago, with the earliest traces of life on Earth emerging by about 3.5 billion years ago.

4. The far future, after cessation of stellar formation, with various scenarios for the ultimate fate of the universe.

Epoch	Time	Redshift	Temperature (Energy)	Description
Planck epoch	$<10^{-43}$ s		$>10^{32}$ K ($>10^{19}$ GeV)	The Planck scale is the scale beyond which current physical theories do not have predictive value. The Planck epoch is the time during which physics is assumed to have been dominated by quantum effects of gravity.
Grand unification epoch	$<10^{-36}$ s		($>10^{16}$ GeV)	The three forces of the Standard Model are unified.

Inflationary epoch, Electroweak epoch	$<10^{-32}$ s		10^{28} K...10^{22} K	Cosmic inflation expands space by a factor of the order of 10^{26} over a time of the order of 10^{-33} to 10^{-32} seconds. The universe is supercooled from about 10^{27} down to 10^{22} kelvins. The Strong Nuclear Force becomes distinct from the Electroweak Force.
Quark epoch	$>10^{-12}$ s		10^{12} K	The forces of the Standard Model have separated, but energies are too high for quarks to coalesce into hadrons, instead forming a quark-gluon plasma. These are the highest energies directly observable in experiment in the Large Hadron Collider.
Hadron epoch	10^{-6} s...1 s		10^{10} K	Quarks are bound into hadrons. A slight matter-antimatter-asymmetry from the earlier phases (baryon asymmetry) results in an elimination of anti-hadrons.
Neutrino decoupling	1 s		10^{10} K (1 MeV)	Neutrinos cease interacting with baryonic matter
Lepton epoch	1 s...10 s		10^{10} K...10^{9} K	Leptons and anti-leptons remain in thermal equilibrium
Photon epoch	10 s...1.2×10^{13} s (...380 ka)		10^{9} K...10^{4} K	The universe consists of a plasma of nuclei, electrons and photons; temperatures remain too high for the binding of electrons to nuclei.
Big Bang nucleosynthesis	10 s...10^{3} s		10^{11} K...10^{9} K (10 MeV...100 keV)	Protons and neutrons are bound into primordial atomic nuclei, hydrogen and helium-4. Small amounts of deuterium, helium-3, and lithium-7 are also synthesized.
Matter-dominated era	47 ka...10 Ga	3600...0.4	10^{4} K...4000 K	During this time, the energy density of matter dominates both radiation density and dark energy, resulting in a decelerated metric expansion of space.
Recombination	380 ka	1100	4000 K	Electrons and atomic nuclei first become bound to form neutral atoms. Photons are no longer in thermal equilibrium with matter and the Universe first becomes transparent. Recombination lasts for about 100 ka, during which Universe is becoming more and more transparent to photons. The photons of the cosmic microwave background radiation originate at this time. The spherical volume of space which will become Observable universe is 42 million light-years in radius at this time.
Dark Ages	380 ka...150 Ma	1100...20	4000 K...60 K	The time between recombination and the formation of the first stars. During this time, the only radiation emitted was the hydrogen line. The chemistry of life may have begun shortly after the Big Bang, 13.8 billion years ago, during a "habitable epoch" when the Universe was only 10-17 million years old.
Stelliferous Era	150 Ma...100 Ga	20...−0.99	60 K...0.03 K	The time between the first formation of Population III stars until the cessation of star formation, leaving all stars in the form of degenerate remnants.

Reionization	150 Ma...1 Ga	20...6	60 K...19 K	The most distant astronomical objects observable with telescopes date to this period; as of 2016, the most remote galaxy observed is GN-z11, at a redshift of 11.09. The earliest "modern" Population III stars are formed in this period.
Galaxy formation and evolution	1 Ga...10 Ga	6...0.4	19 K...4 K	Galaxies coalesce into "proto-clusters" from about 1 Ga (z = 6) and into Galaxy clusters beginning at 3 Gy (z = 2.1), and into superclusters from about 5 Gy (z = 1.2).
Dark-energy-dominated era	>10 Ga	<0.4	<4 K	Matter density falls below dark energy density (vacuum energy), and expansion of space begins to accelerate. This time happens to correspond roughly to the time of the formation of the Solar System and the evolutionary history of life.
Present time	13.8 Ga	0	2.7 K	Farthest observable photons at this moment are CMB photons. They arrive from a sphere with the radius of 46 billion light-years. The spherical volume inside it is commonly referred to as Observable universe.
Far future	>100 Ga	<−0.99	<0.1 K	The Stelliferous Era will end as stars eventually die and fewer are born to replace them, leading to a darkening universe. Various theories suggest a number of subsequent possibilities. Assuming proton decay, matter may eventually evaporate into a Dark Era (heat death). Alternatively the universe may collapse in a Big Crunch. Alternative suggestions include a false vacuum catastrophe or a Big Rip as possible ends to the universe.

Very Early Universe

Planck Epoch

Times shorter than 10^{-43} seconds (Planck time)

The Planck epoch is an era in traditional (non-inflationary) big bang cosmology wherein the temperature was so high that the four fundamental forces—electromagnetism, gravitation, weak nuclear interaction, and strong nuclear interaction—were one fundamental force. Little is understood about physics at this temperature; different hypotheses propose different scenarios. Traditional big bang cosmology predicts a gravitational singularity before this time, but this theory relies on the theory of general relativity, which is thought to break down for this epoch due to quantum effects.

In inflationary cosmology, times before the end of inflation (roughly 10^{-32} second after the Big Bang) does not follow the traditional big bang timeline. Models attempting to formulate processes of the Planck epoch are speculative proposals for "New Physics". Examples include the Hartle–Hawking initial state, string landscape, string gas cosmology, and the ekpyrotic universe.

Grand Unification Epoch

Between 10^{-43} second and 10^{-36} second after the Big Bang

As the universe expanded and cooled, it crossed transition temperatures at which forces separate from each other. These can be regarded as phase transitions much like condensation and freezing phase transitions of ordinary matter. The grand unification epoch began when gravitation separated from the gauge forces. The non-gravitational physics of this epoch would be described by a so-called grand unified theory (GUT). The grand unification epoch ended when the GUT forces further separate into the strong and electroweak forces.

Inflationary Epoch

Before ca. 10^{-32} seconds after the Big Bang

Cosmic inflation was an era of accelerating expansion produced by a hypothesized field called the inflaton, which would have properties similar to the Higgs field and dark energy. While decelerating expansion would magnify deviations from homogeneity, making the universe more chaotic, accelerating expansion would make the universe more homogeneous. A sufficiently long period of inflationary expansion in the past could explain the high degree of homogeneity that is observed in the universe today at large scales, even if the state of the universe before inflation was highly disordered.

Inflation ended when the inflaton field decayed into ordinary particles in a process called "reheating", at which point ordinary Big Bang expansion began. The time of reheating is usually quoted as a time "after the Big Bang". This refers to the time that would have passed in traditional (non-inflationary) cosmology between the Big Bang singularity and the universe dropping to the same temperature that was produced by reheating, even though, in inflationary cosmology, the traditional Big Bang did not occur.

According to the simplest inflationary models, inflation ended at a temperature corresponding to roughly 10^{-32} second after the Big Bang. As explained above, this does not imply that the inflationary era lasted less than 10^{-32} second. In fact, in order to explain the observed homogeneity of the universe, the duration must be longer than 10^{-32} second. In inflationary cosmology, the earliest meaningful time "after the Big Bang" is the time of the end of inflation.

On March 17, 2014, astrophysicists of the BICEP2 collaboration announced the detection of inflationary gravitational waves in the B-mode power spectrum which was interpreted as clear experimental evidence for the theory of inflation. However, on June 19, 2014, lowered confidence in confirming the cosmic inflation findings was reported and finally, on February 2, 2015, a joint analysis of data from BICEP2/Keck and Planck satellite concluded that the statistical "significance [of the data] is too low to be interpreted as a detection of primordial B-modes" and can be attributed mainly to polarized dust in the Milky Way.

Electroweak Epoch

Between 10^{-36} seconds (or the end of inflation) and 10^{-32} seconds after the Big Bang

According to traditional big bang cosmology, the electroweak epoch began 10^{-36} seconds after the Big Bang, when the temperature of the universe was low enough (10^{28} K) to separate the strong force from the electroweak force (the name for the unified forces of electromagnetism and the weak interaction). In inflationary cosmology, the electroweak epoch began when the inflationary epoch ended, at roughly 10^{-32} seconds.

Baryogenesis

There is currently insufficient observational evidence to explain why the universe contains far more baryons than antibaryons. A candidate explanation for this phenomenon must allow the Sakharov conditions to be satisfied at some time after the end of cosmological inflation. While particle physics suggests asymmetries under which these conditions are met, these asymmetries are too small empirically to account for the observed baryon-antibaryon asymmetry of the universe.

Early Universe

After cosmic inflation ends, the universe is filled with a quark–gluon plasma. From this point onwards the physics of the early universe is better understood, and the energies involved in the Quark epoch are directly amenable to experiment.

Supersymmetry Breaking (Speculative)

If supersymmetry is a property of our universe, then it must be broken at an energy that is no lower than 1 TeV, the electroweak symmetry scale. The masses of particles and their superpartners would then no longer be equal, which could explain why no superpartners of known particles have ever been observed.

Electroweak symmetry Breaking and the Quark Epoch

Between 10^{-12} second and 10^{-6} second after the Big Bang

As the universe's temperature falls below a certain very high energy level, it is believed that the Higgs field spontaneously acquires a vacuum expectation value, which breaks electroweak gauge symmetry. This has two related effects:

1. The weak force and electromagnetic force, and their respective bosons (the W and Z bosons and photon) manifest differently in the present universe, with different ranges;

2. Via the Higgs mechanism, all elementary particles interacting with the Higgs field become massive, having been massless at higher energy levels.

At the end of this epoch, the fundamental interactions of gravitation, electromagnetism, the strong interaction and the weak interaction have now taken their present forms, and fundamental particles have mass, but the temperature of the universe is still too high to allow quarks to bind together to form hadrons.

Hadron Epoch

Between 10^{-6} second and 1 second after the Big Bang

The quark–gluon plasma that composes the universe cools until hadrons, including baryons such as protons and neutrons, can form. At approximately 1 second after the Big Bang neutrinos decouple and begin traveling freely through space. This cosmic neutrino background, while unlikely to ever be observed in detail since the neutrino energies are very low, is analogous to the cosmic microwave background that was emitted much later. However, there is strong indirect evidence that the cosmic neutrino background exists, both from Big Bang nucleosynthesis predictions of the helium abundance, and from anisotropies in the cosmic microwave background.

Lepton Epoch

Between 1 second and 10 seconds after the Big Bang

The majority of hadrons and anti-hadrons annihilate each other at the end of the hadron epoch, leaving leptons and anti-leptons dominating the mass of the universe. Approximately 10 seconds after the Big Bang the temperature of the universe falls to the point at which new lepton/anti-lepton pairs are no longer created and most leptons and anti-leptons are eliminated in annihilation reactions, leaving a small residue of leptons.

Photon Epoch

Between 10 seconds and 380,000 years after the Big Bang

After most leptons and anti-leptons are annihilated at the end of the lepton epoch the energy of the universe is dominated by photons. These photons are still interacting frequently with charged protons, electrons and (eventually) nuclei, and continue to do so for the next 380,000 years.

Nucleosynthesis

Between 3 minutes and 20 minutes after the Big Bang

During the photon epoch the temperature of the universe falls to the point where atomic nuclei can begin to form. Protons (hydrogen ions) and neutrons begin to combine into atomic nuclei in the process of nuclear fusion. Free neutrons combine with protons to form deuterium. Deuterium rapidly fuses into helium-4. Nucleosynthesis only lasts for about seventeen minutes, since the temperature and density of the universe has fallen to the point where nuclear fusion cannot continue. By this time, all neutrons have been incorporated into helium nuclei. This leaves about three times more hydrogen than helium-4 (by mass) and only trace quantities of other light nuclei.

Matter Domination

70,000 years after the Big Bang

At this time, the densities of non-relativistic matter (atomic nuclei) and relativistic radiation (photons) are equal. The Jeans length, which determines the smallest structures that can form (due to competition between gravitational attraction and pressure effects), begins to fall and perturbations, instead of being wiped out by free-streaming radiation, can begin to grow in amplitude.

According to ΛCDM, at this stage, cold dark matter dominates, paving the way for gravitational collapse to amplify the tiny inhomogeneities left by cosmic inflation, making dense regions denser and rarefied regions more rarefied. However, because present theories as to the nature of dark matter are inconclusive, there is as yet no consensus as to its origin at earlier times, as currently exist for baryonic matter.

Recombination

ca. 377,000 years after the Big Bang

Hydrogen and helium *atoms* begin to form as the density of the universe falls. This is thought to have occurred about 377,000 years after the Big Bang. Hydrogen and helium are at the beginning ionized, i.e., no electrons are bound to the nuclei, which (containing positively charged protons) are therefore electrically charged (+1 and +2 respectively). As the universe cools down, the electrons get captured by the ions, forming electrically neutral atoms. This process is relatively fast (and faster for the helium than for the hydrogen), and is known as recombination. At the end of recombination, most of the protons in the universe are bound up in neutral atoms. Therefore, the photons' mean free path becomes effectively infinite and the photons can now travel freely: the universe has become transparent. This cosmic event is usually referred to as *decoupling*.

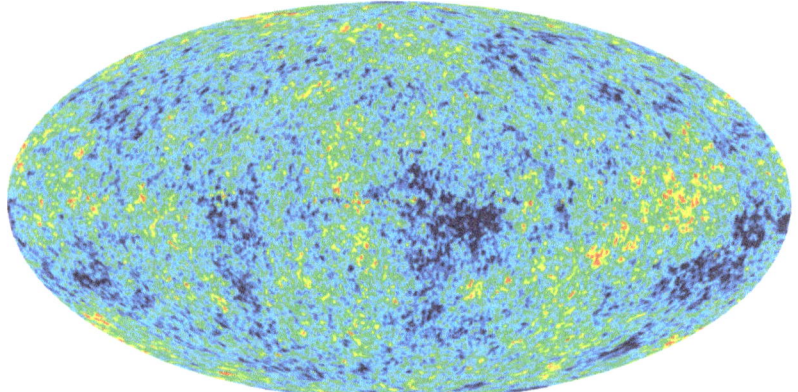

9 year WMAP data (2012) shows the cosmic microwave background radiation variations throughout the universe from our perspective, though the actual variations are much smoother than the diagram suggests.

The photons present at the time of decoupling are the same photons that we see in the cosmic microwave background (CMB) radiation, after being greatly cooled by the expansion of the universe. Around the same time, existing pressure waves within the electron-baryon plasma — known as baryon acoustic oscillations — became embedded in the distribution of matter as it condensed, giving rise to a very slight preference in distribution of large scale objects. Therefore, the cosmic microwave background is a picture of the universe at the end of this epoch including the tiny fluctuations generated during inflation, and the spread of objects such as galaxies in the universe is an indication of the scale and size of the universe as it developed over time.

Dark Ages

ca. 380 thousand – 150 million years after the Big Bang

Before decoupling occurred, most of the photons in the universe were interacting with electrons and protons in the photon–baryon fluid. The universe was opaque or "foggy" as a result. There was light but not light we can now observe through telescopes. The baryonic matter in the universe consisted of ionized plasma, and it only became neutral when it gained free electrons during "recombination", thereby releasing the photons creating the CMB. When the photons were released (or decoupled) the universe became transparent. At this point the only radiation emitted was the 21 cm spin line of neutral hydrogen. There is currently an observational effort underway to detect this faint radiation, as it is in principle an even more powerful tool than the cosmic microwave background for studying the early universe.

The Dark Ages are currently thought to have lasted between 150 million to 800 million years after the Big Bang. The October 2010 discovery of UDFy-38135539, the first observed galaxy to have existed during the following reionization epoch, gives us a window into these times. The galaxy earliest in this period observed and thus also the most distant galaxy ever observed is currently on the record of Leiden University's Richard J. Bouwens and Garth D. Illingsworth from UC Observatories/Lick Observatory. They found the galaxy UDFj-39546284 to be at a time some 480 million years after the Big Bang or about halfway through the Cosmic Dark Ages at a distance of about 13.2 billion light-years. More recently, the UDFy-38135539, EGSY8p7 and GN-z11 galaxies were found to be around 380–550 million years after the Big Bang and at a distance of around 13.4 billion light-years.

Habitable Epoch

ca. 10-17 million years after the Big Bang

The "Dark Ages" span a period during which the temperature of cosmic background radiation cooled from some 4000 K down to about 60 K. The background temperature was between 373 K and 273 K, allowing the possibility of liquid water, during a period of about 6.6 million years, from about 10 to 17 million after the Big Bang (redshift 137–100). Loeb (2014) speculated that primitive life might in principle have appeared during this window, which he called "the Habitable Epoch of the Early Universe".

Large-scale Structure Formation

Structure formation in the big bang model proceeds hierarchically, with smaller structures forming before larger ones. The first structures to form are quasars, which are thought to be bright, early active galaxies, and population III stars. Before this epoch, the evolution of the universe could be understood through linear cosmological perturbation theory: that is, all structures could be understood as small deviations from a perfect homogeneous universe. This is computationally relatively easy to study. At this point non-linear structures begin to form, and the computational problem becomes much more difficult, involving, for example, N-body simulations with billions of particles.

The Hubble Ultra Deep Fields often showcase galaxies from an ancient era that tell us what the early Stelliferous Age was like.

Another Hubble image shows an infant galaxy forming nearby, which means this happened very recently on the cosmological timescale. This shows that new galaxy formation in the universe is still occurring.

Reionization

150 million to 1 billion years after the Big Bang

The first stars and quasars form from gravitational collapse. The intense radiation they emit reionizes the surrounding universe. From this point on, most of the universe is composed of plasma.

Star Formation

The first stars, most likely Population III stars, form and start the process of turning the light elements that were formed in the Big Bang (hydrogen, helium and lithium) into heavier elements. However, as yet there have been no observed Population III stars, and understanding of them is currently based on computational models of their formation and evolution. Fortunately observations of the Cosmic Microwave Background radiation can be used to date when star formation began in earnest. Analysis of such observations made by the European Space Agency's Planck telescope, as reported by BBC News in early February, 2015, concludes that the first generation of stars lit up 560 million years after the Big Bang.

Galaxies, Clusters and Superclusters

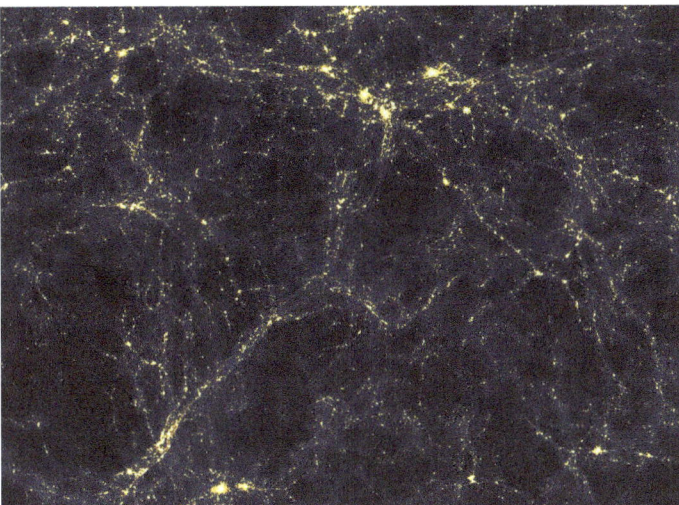

Computer simulated view of the large-scale structure of a part of the universe about 50 million light years across.

Large volumes of matter collapse to form a galaxy. Population II stars are formed early on in this process, with Population I stars formed later.

Johannes Schedler's project has identified a quasar CFHQS 1641+3755 at 12.7 billion light-years away, when the universe was just 7% of its present age.

On July 11, 2007, using the 10-metre Keck II telescope on Mauna Kea, Richard Ellis of the California Institute of Technology at Pasadena and his team found six star forming galaxies about 13.2 billion light years away and therefore created when the universe was only 500 million years old. Only about 10 of these extremely early objects are currently known. More recent observations have shown these ages to be shorter than previously indicated. The most distant galaxy observed as of October 2013 has been reported to be 13.1 billion light years away.

The Hubble Ultra Deep Field shows a number of small galaxies merging to form larger ones, at 13 billion light years, when the universe was only 5% its current age. This age estimate is now believed to be slightly shorter.

Based upon the emerging science of nucleocosmochronology, the Galactic thin disk of the Milky Way is estimated to have been formed 8.8 ± 1.7 billion years ago.

Gravitational attraction pulls galaxies towards each other to form groups, clusters and superclusters.

The Solar System

The Solar System began forming about 4.6 billion years ago, or about 9 billion years after the Big Bang. A fragment of a molecular cloud made mostly of hydrogen and traces of other elements began to collapse, forming a large sphere in the center which would become the Sun, as well as a surrounding disk. The surrounding accretion disk would coalesce into a multitude of smaller objects that would become planets, asteroids, and comets. The Sun is a late-generation star, and the Solar System incorporates matter created by previous generations of stars.

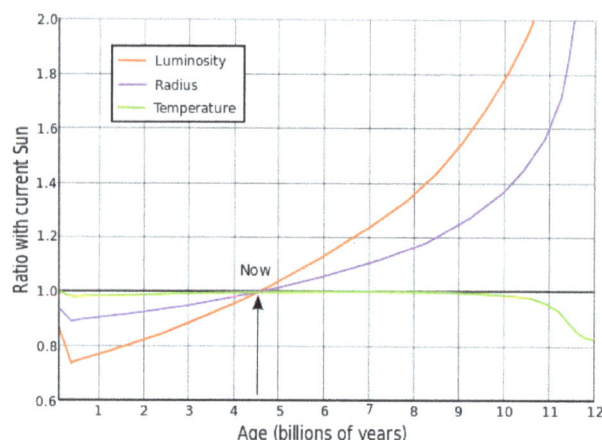

Evolution of the Sun's luminosity, radius and effective temperature compared to the present Sun.

The Big Bang is estimated to have occurred about 13.799 ± 0.021 billion years before present. Since the expansion of the universe appears to be accelerating, its large-scale structure is likely to be the largest structure that will ever form in the universe. The present accelerated expansion prevents any more inflationary structures entering the horizon and prevents new gravitationally bound structures from forming.

The Sun being a main sequence star, its future evolution can be predicted with some certainty. Over a timescale of a billion years or more, the Earth and Solar System are unstable. Earth's existing biosphere is expected to vanish in about a billion years, as the Sun's heat production gradually increases to the point that liquid water and life are unlikely; the Earth's magnetic fields, axial tilt and atmosphere are subject to long-term change; and the Solar System itself is chaotic over million- and billion-year timescales.

Eventually in around 5.4 billion years from now, the core of the Sun will become hot enough to trigger helium fusion in its surrounding shell. This will cause the outer layers of the star to expand greatly, and the star will enter a phase of its life in which it is called a red giant. Within 7.5 billion years, the Sun will have expanded to a radius of 1.2 AU—256 times its current size, and studies announced in 2008 show that due to tidal interaction between Sun and Earth, Earth would actually fall back into a lower orbit, and get engulfed and incorporated inside the Sun before the Sun reaches its largest size, despite the Sun losing about 38% of its mass.

The Sun itself will continue to exist for many billions of years, passing through a number of phases, and eventually ending up as a long-lived white dwarf. Eventually, after billions more years, the Sun will finally cease to shine altogether, becoming a black dwarf.

Ultimate Fate of the Universe

There are several competing scenarios for the possible long-term evolution of the universe. Which of them is going to happen depends on the precise values of physical constants such as the cosmological constant, the possibility of proton decay, and the natural laws beyond the Standard Model.

- Heat Death: In the case of indefinitely continuing metric expansion of space, the energy density in the universe will decrease until, after an estimated time of 10^{1000} years, it reaches thermodynamic equilibrium and no more structure will be possible. This will happen only after an extremely long time because first, all matter will collapse into black holes, which will then evaporate extremely slowly via Hawking radiation. The universe in this scenario will cease to be able to support life much earlier than this, after some 10^{14} years or so, when star formation ceases. In some grand unified theories, proton decay after at least 10^{34} years will convert the remaining interstellar gas and stellar remnants into leptons (such as positrons and electrons) and photons. Some positrons and electrons will then recombine into photons. In this case, the universe has reached a high-entropy state consisting of a bath of particles and low-energy radiation. It is not known however whether it eventually achieves thermodynamic equilibrium. The hypothesis of a universal heat death stems from the 1850s ideas of William Thomson (Lord Kelvin) who extrapolated the theory of heat views of mechanical energy loss in nature, as embodied in the first two laws of thermodynamics, to universal operation.

- Big Rip: For sufficiently large values for the dark energy content of the universe, the expansion rate of the universe will continue to increase without limit. Gravitationally bound systems, such as clusters of galaxies, galaxies, and ultimately the Solar System will be torn apart. Eventually the expansion will be so rapid as to overcome the electromagnetic forces holding molecules and atoms together. Finally even atomic nuclei will be torn apart and the universe as we know it will end in an unusual kind of gravitational singularity.

- Big Crunch: In the opposite of the "Big Rip" scenario, the metric expansion of space would at some point be reversed and the universe would contract towards a hot, dense state. This is a required element of oscillatory universe scenarios, such as the cyclic model, although a Big Crunch does not necessarily imply an oscillatory universe. Current observations suggest that this model of the universe is unlikely to be correct, and the expansion will continue or even accelerate.

- Vacuum instability: Cosmology traditionally has assumed a stable or at least metastable universe, but the possibility of a false vacuum in quantum field theory implies that the universe at any point in spacetime might spontaneously collapse into a lower energy state, a more stable or "true vacuum", which would then expand outward from that point with the speed of light.

References

- Christopher J. Conselice; et al. (2016). "The Evolution of Galaxy Number Density at z < 8 and its Implications". The Astrophysical Journal. 830 (2): 83. arXiv:1607.03909v2. doi:10.3847/0004-637X/830/2/83

- Itzhak Bars; John Terning (November 2009). Extra Dimensions in Space and Time. Springer. pp. 27–. ISBN 978-0-387-77637-8. Retrieved 2011-05-01

- "Seven-Year Wilson Microwave Anisotropy Probe (WMAP) Observations: Sky Maps, Systematic Errors, and Basic Results" (PDF). nasa.gov. Retrieved 2010-12-02

- M. J. Geller; J. P. Huchra (1989). "Mapping the universe.". Science. 246 (4932): 897–903. Bibcode:1989Sci...246..897G. PMID 17812575. doi:10.1126/science.246.4932.897

- Fixsen, D. J. (December 2009). "The Temperature of the Cosmic Microwave Background". The Astrophysical Journal. 707 (2): 916–920. Bibcode:2009ApJ...707..916F. arXiv:0911.1955. doi:10.1088/0004-637X/707/2/916

- Carroll, Bradley W.; Ostlie, Dale A. (2013-07-23). An Introduction to Modern Astrophysics (International ed.). Pearson. p. 1178. ISBN 9781292022932

- Dreifus, Claudia (2 December 2014). "Much-Discussed Views That Go Way Back - Avi Loeb Ponders the Early Universe, Nature and Life". New York Times. Retrieved 3 December 2014

- Planck collaboration (2013). "Planck 2013 results. XVI. Cosmological parameters". Astronomy & Astrophysics. 571: A16. Bibcode:2014A&A...571A..16P. arXiv:1303.5076. doi:10.1051/0004-6361/201321591

- Loeb, Abraham (October 2014). "The Habitable Epoch of the Early Universe". International Journal of Astrobiology. 13 (04): 337–339. Bibcode:2014IJAsB..13..337L. doi:10.1017/S1473550414000196. Retrieved 15 December 2014

- Robert P Kirshner (2002). The Extravagant Universe: Exploding Stars, Dark Energy and the Accelerating Cosmos. Princeton University Press. p. 71. ISBN 0-691-05862-8

- Rampelotto, P.H. (2010). "Panspermia: A Promising Field Of Research" (PDF). Astrobiology Science Conference. Retrieved 3 December 2014

- John L Tonry; et al. (2003). "Cosmological Results from High-z Supernovae". Astrophys J. 594: 1. Bibcode:2003ApJ...594....1T. arXiv:astro-ph/0305008. doi:10.1086/376865

- del Peloso, E. F. (2005). "The age of the Galactic thin disk from Th/Eu nucleocosmochronology. III. Extended sample". Astronomy and Astrophysics. 440 (3): 1153–1159. Bibcode:2005A&A...440.1153D. arXiv:astro-ph/0506458. doi:10.1051/0004-6361:20053307

- Massimo Giovannini (2008). A primer on the physics of the cosmic microwave background. World Scientific. pp. 70–. ISBN 978-981-279-142-9. Retrieved 1 May 2011

- Lineweaver, Charles; Tamara M. Davis (2005). "Misconceptions about the Big Bang" (PDF). Scientific American. Retrieved 2008-11-06

- Coleman, Sidney; De Luccia, Frank (1980-06-15). "Gravitational effects on and of vacuum decay" (PDF). Physical Review D. D21 (12): 3305–3315. Bibcode:1980PhRvD..21.3305C. doi:10.1103/PhysRevD.21.3305

- Overbye, Dennis (March 17, 2014). "Space Ripples Reveal Big Bang's Smoking Gun". The New York Times. Retrieved March 17, 2014

- R. Brent Tully; Helene Courtois; Yehuda Hoffman; Daniel Pomarède (2 September 2014). "The Laniakea supercluster of galaxies". Nature (published 4 September 2014). 513 (7516): 71. Bibcode:2014Natur.513...71T. PMID 25186900. arXiv:1409.0880. doi:10.1038/nature13674

Understanding Religious Cosmology

Religious cosmology is the explanation of the universe from a religious point of view. It studies the religious interpretation of the origin and history of the cosmos. The topics discussed in the chapter are of great importance to broaden the existing knowledge on cosmology.

Religious Cosmology

A religious cosmology (also mythological cosmology) is a way of explaining the origin, the history and the evolution of the cosmos or universe based on the religious mythology of a specific tradition. Religious cosmologies usually include an act or process of creation by a creator deity or a larger pantheon.

Biblical Cosmology

The universe of the ancient Israelites was made up of a flat disc-shaped earth floating on water, heaven above, underworld below. Humans inhabited earth during life and the underworld after death, and the underworld was morally neutral; only in Hellenistic times (after c.330 BC) did Jews begin to adopt the Greek idea that it would be a place of punishment for misdeeds, and that the righteous would enjoy an afterlife in heaven. In this period too the older three-level cosmology was widely replaced by the Greek concept of a spherical earth suspended in space at the centre of a number of concentric heavens.

Christianity/Modern Judaism

Around the time of Jesus or a little earlier, the Greek idea that God had actually created matter replaced the older idea that matter had always existed, but in a chaotic state. This concept, called *creatio ex nihilo*, is now the accepted orthodoxy of most denominations of Judaism and Christianity. Most denominations of Christianity and Judaism believe that a single, uncreated God was responsible for the creation of the cosmos.

Mormon Cosmology

The Earth's creation, according to Mormon scripture, was not *ex nihilo*, but organized from existing matter. The faith teaches that this earth is just one of many inhabited worlds, and that there are many governing heavenly bodies, including a planet or star Kolob which is said to be nearest the throne of God. According to the King Follett discourse, God the Father himself once passed through mortality like Jesus did, but how, when, or where that took place is unclear. The prevailing view among Mormons is that God once lived on a planet.

Buddhism

In Buddhism, the universe comes into existence dependent upon the actions (karma) of its inhabitants. Buddhists posit neither an ultimate beginning nor final end to the universe, but see the universe as something in flux, passing in and out of existence, parallel to an infinite number of other universes doing the same thing.

The Buddhist universe consists of a large number of worlds which correspond to different mental states, including passive states of trance, passionless states of purity, and lower states of desire, anger, and fear. The beings in these worlds are all coming into existence or being born, and passing out of existence into other states, or dying. A world comes into existence when the first being in it is born, and ceases to exist, as such, when the last being in it dies. The universe of these worlds also is born and dies, with the death of the last being preceding a universal conflagration that destroys the physical structure of the worlds; then, after an interval, beings begin to be born again and the universe is once again built up. Other universes, however, also exist, and there are higher planes of existence which are never destroyed, though beings that live in them also come into and pass out of existence.

As well as a model of universal origins and destruction, Buddhist cosmology also functions as a model of the mind, with its thoughts coming into existence based on preceding thoughts, and being transformed into other thoughts and other states.

Islam

Map of the world according to Zakariya al-Qazwini showing his view of how the universe is structured and how sky and earth are supported. The Earth is considered flat and surrounded by a series of mountains—including Mount Qaf—that hold it in its place like pegs; the Earth is supported by an ox that stands on Bahamut dwelling in a cosmic ocean; the ocean is inside a bowl that sits on top of an angel or jinn.

Islam teaches that God created the universe, including Earth's physical environment and human beings. The highest goal is to visualize the cosmos as a book of symbols for meditation and contemplation for spiritual upliftment or as a prison from which the human soul must escape to attain true freedom in the spiritual journey to God.

Below here there are some other citations from the Quran on cosmology.

"And the heavens We constructed with strength, and indeed, We are [its] expander." 51:47 Sahih International

"Do not the unbelievers see that the heavens and the earth were joined together (as one unit of creation), before We clove them asunder? We made from water every living thing. Will they not then believe?" 21:30 Yusuf Ali translation

"The day that We roll up the heavens like a scroll rolled up for books (completed),- even as We produced the first creation, so shall We produce a new one: a promise We have undertaken: truly shall We fulfil it." 21:104 Yusuf Ali translation

Hinduism

The Hindu cosmology indicates that the present cycle is not the beginning of everything but preceded by an infinite number of universes and to be followed by another infinite number of universes.

The Rig Veda questions the origin of the cosmos in: "Neither being (sat) nor non-being was as yet. What was concealed? And where? And in whose protection?...Who really knows? Who can declare it? Whence was it born, and whence came this creation? The devas (demigods) were born later than this world's creation, so who knows from where it came into existence? None can know from where creation has arisen, and whether he has or has not produced it. He who surveys it in the highest heavens, he alone knows-or perhaps does not know."

The Rig Veda's view of the cosmos also sees one true divine principle self-projecting as the divine word, *Vaak*, 'birthing' the cosmos that we know, from the monistic *Hiranyagarbha* or Golden Womb. The *Hiranyagarbha* is alternatively viewed as Brahma, the creator who was in turn created by God, or as God (Brahman) himself. The creation begins anew after billions of years (Solar years) of non-existence.

Brahma's day is divided in one thousand cycles *(Maha Yuga*, or the Great Year). *Maha Yuga*, during which life, including the human race appears and then disappears, has 71 divisions, each made of 14 *Manvantara* (1000) years. Each *Maha Yuga* lasts for 4,320,000 years. *Manvantara* is Manu's cycle, the one who gives birth and governs the human race.

Each *Maha Yuga* consists of a series of four shorter *yugas*, or ages. The *yugas* get progressively worse from a moral point of view as one proceeds from one *yuga* to another. As a result, each *yuga* is of shorter duration than the age that preceded it. The current *Kali Yuga* (Iron Age) began at midnight 17 February / 18 February in 3102 BC in the proleptic Julian calendar.

Jainism

Jain cosmology considers the loka, or universe, as an uncreated entity, existing since infinity, having no beginning or an end. Jain texts describe the shape of the universe as similar to a man standing with legs apart and arm resting on his waist. This Universe, according to Jainism, is narrow at the top, broad at the middle and once again becomes broad at the bottom.

Mahāpurāṇa of Ācārya Jinasena is famous for this quote: "Some foolish men declare that a creator made the world. The doctrine that the world was created is ill advised and should be rejected. If God created the world, where was he before the creation? If you say he was transcendent then and needed no support, where is he now? How could God have made this world without any raw material? If you say that he made this first, and then the world, you are faced with an endless regression."

Chinese Mythology

There is a "primordial universe" Wuji (philosophy), and Hongjun Laozu, water or qi. It transformed into Taiji and multiplied into everything. The Pangu legend tells a formless chaos coalesced into a cosmic egg. Pangu emerged (or woke up) and separated Yin from Yang with a swing of his giant axe, creating the Earth (murky *Yin*) and the Sky (clear *Yang*). To keep them separated, Pangu stood between them and pushed up the Sky. After Pangu died, he became everything.

Jain Cosmology

Jain cosmology is the description of the shape and functioning of the Universe (*loka*) and its constituents (such as living beings, matter, space, time etc.) according to Jainism. Jain cosmology considers the universe, as an uncreated entity, existing since infinity, having neither beginning nor end. Jain texts describe the shape of the universe as similar to a man standing with legs apart and arm resting on his waist. This Universe, according to Jainism, is broad at the top, narrow at the middle and once again becomes broad at the bottom.

Six Eternal Substances

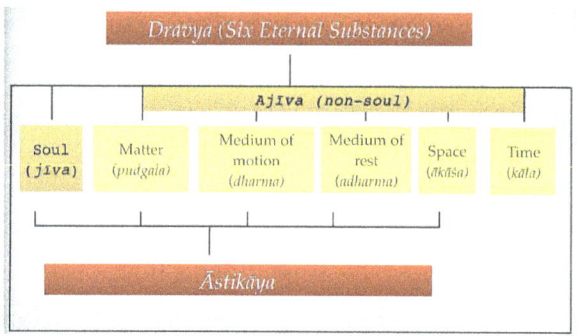

Chart showing the classification of *dravya* and *astikaya*

According to Jains, the Universe is made up of six simple and eternal substances called *dravya* classified as follows:

- *Jīva* (Living Substances)

  Jīva (Jainism)|Jīva i.e. Souls - *Jīva* exists as a reality, having a separate existence from the body that houses it. It is characterised by *chetana* (consciousness) and *upayoga* (knowledge and perception). Though the soul experiences both birth and death, it is neither really destroyed nor created. Decay and origin refer respectively to the disappearing of one state of soul and appearing of another state, these being merely the modes of the soul.

- *Ajīva* (Non-Living Substances)

- *Pudgala* (Matter) - Matter is classified as solid, liquid, gaseous, energy, fine Karmic materials and extra-fine matter i.e. ultimate particles. *Paramān u* or ultimate particle is the basic building block of all matter. The Paramāṇu and Pudgala are permanent and indestructible. Matter combines and changes its modes but its basic qualities remain the same. According to Jainism, it cannot be created, nor destroyed.

- *Dharma-dravya* (Principle of Motion) and

- *Adharma-dravya* (Principle of Rest) - *Dharmastikāya* and *Adharmastikāya* are distinctly peculiar to Jaina system of thought depicting the principle of Motion and Rest. They are said to pervade the entire universe. *Dharma* and *Adharma* are by itself not motion or rest but mediate motion and rest in other bodies. Without Dharmastikāya motion is not possible and without Adharmastikāya rest is not possible in the universe.

- Ākāśa (Space) - Space is a substance that accommodates the living souls, the matter, the principle of motion, the principle of rest and time. It is all-pervading, infinite and made of infinite space-points.

- *Kāla* (Time) - *Kāla* is an eternal substance according to Jainism and all activities, changes or modifications can be achieved only through the progress of time. According to the Jain text, Dravyasaṃgraha:

Conventional time (*vyavahāra kāla*) is perceived by the senses through the transformations and modifications of substances. Real time (*niścaya kāla*), however, is the cause of imperceptible, minute changes (called *vartanā*) that go on incessantly in all substances.

—*Dravyasamgraha (21)*

Universe and its Structure

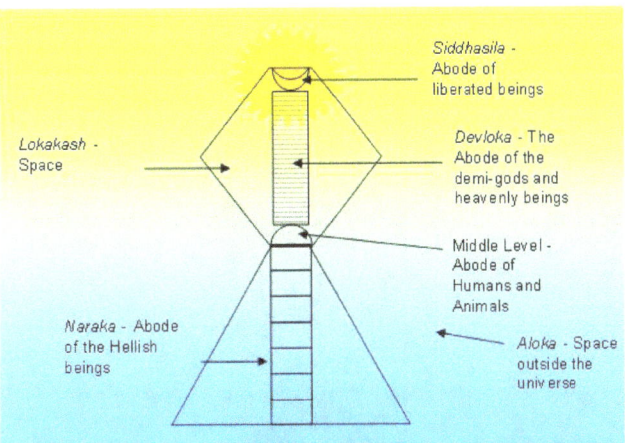

Structure of Universe according to the Jain scriptures.

The Jain doctrine postulates an eternal and ever-existing world which works on universal natural laws. The existence of a creator deity is overwhelmingly opposed in the Jain doctrine. Mahāpurāṇa, a Jain text authored by *Ācārya Jinasena* is famous for this quote:

" Some foolish men declare that a creator made the world. The doctrine that the world was created is ill advised and should be rejected. If God created the world, where was he before the creation? If you say he was transcendent then and needed no support, where is he now? How could God have made this world without any raw material? If you say that he made this first, and then the world, you are faced with an endless regression. "

According to Jains, the universe has a firm and an unalterable shape which is measured in the Jain texts by means of a unit called *Rajju* which is supposed to be very large. The Digambara sect of Jainism postulates that the universe is fourteen Rajju high and extends seven Rajjus from north to south. Its breadth is seven Rajjus at the bottom and decreases gradually till the middle where it is one Rajju. The width then increases gradually till it is five Rajju and again decreases till it is one Rajju. The apex of the universe is one Rajju long, one Rajju wide and eight Rajju high. The total space of the world is thus 343 cubic Rajju. The svetambara view differs slightly and postulates that there is constant increase and decrease in the breadth and the space is 239 cubic Rajju. Apart from the apex which is the abode of liberated beings, the universe is divided into three parts. The world is surrounded by three atmospheres: dense-water, dense-wind and thin-wind. It is then surrounded by infinitely large non-world which is absolutely empty.

The whole world is said to be filled with living beings. In all the three parts, there is the existence of very small living beings called nigoda. Nigoda are of two types: nitya-nigoda and Itara-nigoda. Nitya-nigoda are those which will reborn as nigoda throughout eternity where as Itara-nigoda will be reborn as other beings too. The mobile region of universe (Trasandi) is one Rajju wide, one Rajju broad and fourteen Rajju high. Within this, there are animals and plants everywhere where as Human beings are restricted to 2.5 continents of middle world. The beings inhabiting lower world are called Naraki (Hellish beings). Deva (roughly demi-gods) live in whole of the top and middle world and top three realms of lower world. Living beings are divided in fourteen classes (Jivasthana) : 1. fine beings with one sense. 2. Crude beings with one sense. 3. beings with two sense. 4. beings with three sense. 5. Beings with four sense. 6. beings with five sense without mind. 7. beings with five sense with a mind. These can be under-developed or developed which makes it a total of fourteen. Human beings get any form of existence and are the only ones which can attain salvation.

Three Lokas

Fourteen Rajaloka or Triloka. Shape of Universe as per Jain cosmology in form of a cosmic man.

Miniature from 17th century, *Saṁgrahaṇīratna* by Śrīcandra, in Prakrit with a Gujarati commentary. Jain Śvetāmbara cosmological text with commentary and illustrations.

The early Jains contemplated the nature of the earth and universe and developed a detailed hypothesis on the various aspects of astronomy and cosmology. According to the Jain texts, the universe is divided into 3 parts:

- *Urdhva Loka* – the realms of the gods or heavens

- *Madhya Loka* – the realms of the humans, animals and plants

- *Adho Loka* – the realms of the hellish beings or the infernal regions

The following Upanga āgamas describe the Jain cosmology and geography in a great detail:

1. Sūryaprajñapti – Treatise on Sun

2. Jambūdvīpaprajñapti - Treatise on the island of Roseapple tree; it contains a description of Jambūdvī and life biographies of Ṛṣabha and King Bharata

3. Candraprajñapti - Treatise on moon

Additionally, the following texts describe the Jain cosmology and related topics in detail:

1. Trilokasāra – Essence of the three worlds (heavens, middle level, hells)

2. Trilokaprajñapti – Treatise on the three worlds

3. Trilokadipikā – Illumination of the three worlds

4. Tattvārthasūtra - Description on nature of realities

5. Kṣetrasamasa – Summary of Jain geography

6. Bruhatsamgrahni – Treatise on Jain cosmology and geography

Urdhva Loka, the Upper World

Upper World (Udharva loka) is divided into different abodes and are the realms of the heavenly beings (demi-gods) who are non-liberated souls.

Upper World is divided into sixteen Devalokas, nine Graiveyaka, nine Anudish and five Anuttar abodes. Sixteen Devaloka abodes are Saudharma, Aishana, Sanatkumara, Mahendra, Brahma, Brahmottara, Lantava, Kapishta, Shukra, Mahashukra, Shatara, Sahasrara, Anata, Pranata, Arana and Achyuta. Nine Graiveyak abodes are Sudarshan, Amogh, Suprabuddha, Yashodhar, Subhadra, Suvishal, Sumanas, Saumanas and Pritikar. Nine Anudish are Aditya, Archi, Archimalini, Vair, Vairochan, Saum, Saumrup, Ark and Sphatik. Five Anuttar are Vijaya, Vaijayanta, Jayanta, Aparajita and Sarvarthasiddhi.

The sixteen heavens in Devalokas are also called Kalpas and the rest are called Kalpatit. Those living in Kalpatit are called Ahamindra and are equal in grandeur. There is increase with regard to the lifetime, influence of power, happiness, lumination of body, purity in thought-colouration, capacity of the senses and range of clairvoyance in the Heavenly beings residing in the higher abodes. But there is decrease with regard to motion, stature, attachment and pride. The higher groups, dwelling in 9 Greveyak and 5 Anutar Viman. They are independent and dwelling in their own vehicles. The anuttara souls attain liberation within one or two lifetimes. The lower groups, organized like earthly kingdoms—rulers (Indra), counselors, guards, queens, followers, armies etc.

Above the Anutar vimans, at the apex of the universe, is the Siddhasila, the realms of the liberated souls also known as the Siddhas, the perfected omniscient and blissful beings, who are venerated by the Jains.

Madhya Loka, the Middle World

Image depicting map of Jambudvipa as per Jain Cosmology

Early 19th-century painting depicting map of 2 $\frac{1}{2}$ continents

Depiction of Mount Meru at Jambudweep, Hastinapur

Madhya Loka, at the centre of the universe consists of 900 yojans above and 900 yojans below earth surface. It is inhabited by:

1. *Jyotishka devas* (luminous gods) - 790 to 900 yojans above earth

2. Humans, Tiryanch (Animals, birds, plants) on the surface

3. *Vyantar devas* (Intermediary gods)- 100 yojan below the ground level

Madhyaloka consists of many continent-islands surrounded by oceans, first eight whose names are:

Continent/ Island	Ocean
Jambūdvīpa	*Lavanoda (Salt - ocean)*
Ghatki Khand	*Kaloda (Black sea)*
Puskarvardvīpa	*Puskaroda (Lotus Ocean)*
Varunvardvīpa	*Varunoda (Varun Ocean)*
Kshirvardvīpa	*Kshiroda (Ocean of milk)*
Ghrutvardvīpa	*Ghrutoda (Butter milk ocean)*
Ikshuvardvīpa	*Iksuvaroda (Sugar Ocean)*
Nandishwardvīpa	*Nandishwaroda*

Mount Meru (also *Sumeru*) is at the centre of the world surrounded by Jambūdvīpa, in form of a circle forming a diameter of 100,000 yojans. There are two sets of sun, moon and stars revolving around Mount Meru; while one set works, the other set rests behind the Mount Meru.

MS 4465
Sricandra: Sankhitta sangheyani; the concise compendium of cosmography. West India, 17th c.

Work of Art showing maps and diagrams as per Jain Cosmography from 17th century
CE Manuscript of 12th century Jain text *Sankhitta Sangheyan*

Jambūdvīpa continent has 6 mighty mountains, dividing the continent into 7 zones (Ksetra). The names of these zones are:

1. Bharat Kshetra

2. Mahavideh Kshetra

3. Airavat Kshetra

4. Ramyak

5. Hairanyvat Kshetra

6. Haimava Kshetra

7. Hari Kshetra

The three zones i.e. Bharat Kshetra, Mahavideh Kshetra and Airavat Kshetra are also known as Karma bhoomi because practice of austerities and liberation is possible and the Tirthankaras preach the Jain doctrine. The other four zones, Ramyak, Hairanyvat Kshetra, Haimava Kshetra and Hari Kshetra are known as akarmabhoomi or bhogbhumi as humans live a sinless life of pleasure and no religion or liberation is possible.

Nandishvara Dvipa is not the edge of cosmos, but it is beyond the reach of humans. Humans can reside only on *Jambudvipa*, *Dhatatikhanda Dvipa*, and the inner half of *Pushkara Dvipa*.

Adho Loka, the Lower World

17th century cloth painting depicting seven levels of Jain hell and various tortures suffered in them. Left panel depicts the demi-god and his animal vehicle presiding over the each hell.

The lower world consists of seven hells, which are inhabited by Bhavanpati demigods and the hellish beings. Hellish beings reside in the following hells -

1. Ratna prabha-dharma.

2. Sharkara prabha-vansha.

3. Valuka prabha-megha.

4. Pank prabha-anjana.

5. Dhum prabha-arista.

6. Tamah prabha-maghavi.

7. Mahatamah prabha-maadhavi

Time Cycle

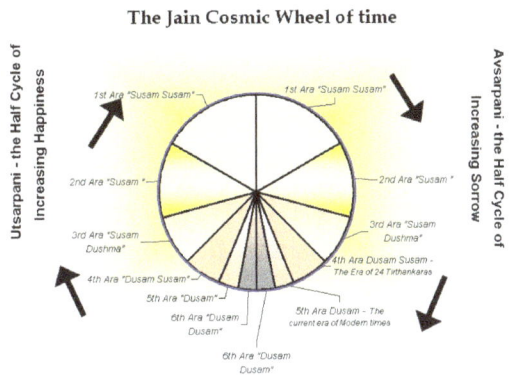

The Jain Cosmic Wheel of time

Division of time as envisaged by Jains

According to Jainism, time is beginningless and eternal. The *Kālacakra*, the cosmic wheel of time, rotates ceaselessly. The wheel of time is divided into two half-rotations, *Utsarpiṇī* or ascending time cycle and *Avasarpiṇī*, the descending time cycle, occurring continuously after each other. *Utsarpiṇī* is a period of progressive prosperity and happiness where the time spans and ages are at an increasing scale, while *Avsarpiṇī* is a period of increasing sorrow and immorality with decline in timespans of the epochs. Each of this half time cycle consisting of innumerable period of time (measured in *sagaropama* and *palyopama* years) is further sub-divided into six *aras* or epochs of unequal periods. Currently, the time cycle is in *avasarpiṇī* or descending phase with the following epochs.

Name of the Ara	Degree of happiness	Duration of Ara	Maximum height of people	Maximum lifespan of people
Suṣama-suṣamā	Utmost happiness and no sorrow	400 trillion sāgaropamas	Six miles tall	Three Palyopam years
Suṣamā	Moderate happiness and no sorrow	300 trillion sāgaropamas	Four miles tall	Two Palyopam Years
Suṣama-duḥṣamā	Happiness with very little sorrow	200 trillion sāgaropamas	Two miles tall	One Palyopam Years
Duḥṣama-suṣamā	Happiness with little sorrow	100 trillion sāgaropamas	1500 meters	84 Lakh Purva
Duḥṣamā	Sorrow with very little happiness	21,000 years	7 hatha	120 years
Duḥṣama-duḥṣamā	Extreme sorrow and misery	21,000 years	1 hatha	20 years

In *utsarpiṇī* the order of the eras is reversed. Starting from *duṣamā-duṣamā*, it ends with *suṣamā-suṣamā* and thus this never ending cycle continues. Each of these aras progress into the next phase seamlessly without any apocalyptic consequences. The increase or decrease in the happiness, life spans and length of people and general moral conduct of the society changes in a phased and graded manner as the time passes. No divine or supernatural beings are credited or responsible with these spontaneous temporal changes, either in a creative or overseeing role, rather human beings and creatures are born under the impulse of their own *karmas*.

Śalākāpuruṣas - The Deeds of the 63 Illustrious Men

During each motion of the half-cycle of the wheel of time, 63 *Śalākāpuruṣa* or 63 illustrious men, consisting of the 24 *Tīrthaṅkaras* and their contemporaries regularly appear. The Jain universal or legendary history is basically a compilation of the deeds of these illustrious men. They are categorised as follows:

- 24 Tīrthaṅkaras – The 24 Tīrthaṅkaras or the ford makers appear in succession to activate the true religion and establish the community of ascetics and laymen.

- 12 Chakravartins – The Chakravartīs are the universal monarchs who rule over the six continents.

- 9 Balabhadras who lead an ideal Jain life.

- 9 *Narayana* or *Vasudev* (heroes)

- 9 *Prati-Naryana* or *Prati-Vasudev* (anti-heroes) – They are anti-heroes who are ultimately killed by the *Narayana*.

Balabhadra and *Narayana* are half brothers who jointly rule over three continents.

Besides these a few other important classes of 106 persons are recognized:-

- 9 Naradas

- 11 Rudras

- 24 Kamdevas

- 24 Fathers of the Tirthankaras.

- 24 Mothers of the Tirthankaras.

- 14 *Kulakara* (patriarchs)

Buddhist Cosmology

Buddhist cosmology is the description of the shape and evolution of the Universe according to the Buddhist scriptures and commentaries.

It consists of temporal and spatial cosmology, the temporal cosmology being the division of the existence of a 'world' into four discrete moments (the creation, duration, dissolution, and state of being dissolved, this does not seem to be a canonical division however). The spatial cosmology consists of a vertical cosmology, the various planes of beings, their bodies, characteristics, food, lifespan, beauty and a horizontal cosmology, the distribution of these world-systems into an "apparently" infinite sheet of universes. The existence of world-periods (moments, kalpas), is well attested to by the Buddha.

The historical Buddha (Gautama Buddha) made references to the existence of aeons (which he describes the length of by metaphor), and simultaneously intimates his knowledge of past events, such as the dawn of human beings in their coarse and gender-split forms, the existence of there being more than one sun at certain points in time, and his ability to convey his voice vast distances, as well as the ability of his disciples (who if they fare accordingly) to be reborn in any one of these planes (should they so choose)—the Buddha does not seem to place a premium on figuring out cosmology.

He also refused to answer questions regarding either the infinitude or eternity of the world.

Introduction

The self-consistent Buddhist cosmology which is presented in commentaries and works of Abhidharma in both Theravāda (31 planes) and Mahāyāna traditions, is the end-product of an analysis and reconciliation of cosmological comments found in the Buddhist sūtra and vinaya traditions. No single sūtra sets out the entire structure of the universe. Kalpa Vibhangaya However, in several sūtras the Buddha describes other worlds and states of being, and other sūtras describe the origin and destruction of the universe. The synthesis of these data into a single comprehensive system must have taken place early in the history of Buddhism, as the system described in the Pāli Vibhajyavāda tradition (represented by today's Theravādins) agrees, despite some minor inconsistencies of nomenclature, with the Sarvāstivāda tradition which is preserved by Mahāyāna Buddhists.

The picture of the world presented in Buddhist cosmological descriptions cannot be taken as a literal description of the shape of the universe. It is inconsistent, and cannot be made consistent, with astronomical data that were already known in ancient India. However, it is not intended to be a description of how ordinary humans perceive their world; rather, it is the universe as seen through the divyacak us (Pāli: dibbacakkhu), the "divine eye" by which a Buddha or an arhat who has cultivated this faculty can perceive all of the other worlds and the beings arising (being born) and passing away (dying) within them, and can tell from what state they have been reborn and into what state they will be reborn. The cosmology has also been interpreted in a symbolical or allegorical sense.

Buddhist cosmology can be divided into two related kinds: spatial cosmology, which describes the arrangement of the various worlds within the universe; and temporal cosmology, which describes how those worlds come into existence, and how they pass away.

Spatial Cosmology

Spatial cosmology can also be divided into two branches. The *vertical* (or cakravāla) cosmology describes the arrangement of worlds in a vertical pattern, some being higher and some lower. By contrast, the *horizontal* (sahasra) cosmology describes the grouping of these vertical worlds into sets of thousands, millions or billions.

Vertical Cosmology

In the vertical cosmology, the universe exists of many worlds (lokāḥ) – one might say "planes/ realms" – stacked one upon the next in layers. Each world corresponds to a mental state or a state of being. A world is not, however, a location so much as it is the beings which compose it; it is sustained by their karma and if the beings in a world all die or disappear, the world disappears too. Likewise, a world comes into existence when the first being is born into it. The physical separation is not so important as the difference in mental state; humans and animals, though they partially share the same physical environments, still belong to different worlds because their minds perceive and react to those environments differently.

The vertical cosmology is divided into thirty-one planes of existence and the planes into three realms, or dhātus, each corresponding to a different type of mentality. These three realms (Tridhātu) are the Ārūpyadhātu (4 Realms), the Rūpadhātu (16 Realms), and the Kāmadhātu (15 Realms). This Sakwala/solar system or plane of existence comprises the "five or six desire realms". In some instances all of the beings born in the Ārūpyadhātu and the Rūpadhātu are informally classified as "gods" or "deities" (devāḥ), along with the gods of the Kāmadhātu, notwithstanding the fact that the deities of the Kāmadhātu differ more from those of the Ārūpyadhātu than they do from humans. It is to be understood that deva is an imprecise term referring to any being living in a longer-lived and generally more blissful state than humans. Most of them are not "gods" in the common sense of the term, having little or no concern with the human world and rarely if ever interacting with it; only the lowest deities of the Kāmadhātu correspond to the gods described in many polytheistic religions.

The term "brahmā" is used both as a name and as a generic term for one of the higher devas. In its broadest sense, it can refer to any of the inhabitants of the Ārūpyadhātu and the Rūpadhātu. In more restricted senses, it can refer to an inhabitant of one of the eleven lower worlds of the Rūpadhātu, or in its narrowest sense, to the three lowest worlds of the Rūpadhātu (Plane of Brahma's retinue) A large number of devas use the name "Brahmā", e.g. Brahmā Sahampati, Brahmā Sanatkumāra, Baka Brahmā, etc. It is not always clear which world they belong to, although it must always be one of the worlds of the Rūpadhātu. According to the Ayacana Sutta, Brahmā Sahampati, who begs the Buddha to teach Dhamma to the world, resides in the Śuddhāvāsa worlds.

Formless Realm (Ārūpyadhātu)

The Ārūpyadhātu (Sanskrit) or Arūpaloka (Pāli) or "Formless realm" would have no place in a purely physical cosmology, as none of the beings inhabiting it has either shape or location; and correspondingly, the realm has no location either. This realm belongs to those devas who attained and remained in the Four Formless Absorptions (catuḥ-samāpatti) of the arūpadhyānas in a previous life, and now enjoys the fruits (vipāka) of the good karma of that accomplishment. Bodhisattvas, however, are never born in the Ārūpyadhātu even when they have attained the arūpadhyānas.

There are four types of Ārūpyadhātu devas, corresponding to the four types of arūpadhyānas:

Arupa Bhumi (Arupachara Brahmalokas or Immaterial/Formless Brahma Realms)

- Naivasamjñānāsamjñāyatana or Nevasaññānāsaññāyatana "Sphere of neither perception nor non-perception". In this sphere the formless beings have gone beyond a mere negation of perception and have attained a liminal state where they do not engage in "perception" (samjñā, recognition of particulars by their marks) but are not wholly unconscious. This was the sphere reached by Udraka Rāmaputra (Pāli: Uddaka Rāmaputta), the second of the Buddha's original teachers, who considered it equivalent to enlightenment. Total life span on this realm in human years - 84,000 Maha Kalpa (Maha Kalpa = 4 Asankya Kalpa).

This realm is placed 5,580,000 Yodun (1 Yoduna = 16 Miles) above the Plane of Nothingness(Akiknchaknkayatana).

- Ākiṃcanyāyatana or Ākiñcaññāyatana "Sphere of Nothingness" (literally "lacking anything"). In this sphere formless beings dwell contemplating upon the thought that "there is no thing". This is considered a form of perception, though a very subtle one. This was the sphere reached by Ārāḍa Kālāma, the first of the Buddha's original teachers; he considered it to be equivalent to enlightenment. Total life span on this realm in human years - 60,000 Maha Kalpa. This realm is placed 5,580,000 Yodun above the Plane of Infinite Consciousness(Viknknanaknchayathana).

- Vijñānānantyāyatana or Viññāṇānañcāyatana or more commonly the contracted form Viññāṇañcāyatana "Sphere of Infinite Consciousness". In this sphere formless beings dwell meditating on their consciousness (vijñāna) as infinitely pervasive. Total life span on this realm in human years - 40,000 Maha Kalpa. This realm is placed 5,580,000 Yodun above the Plane of Infinite Space(Akasanknayathanaya)

- Ākāśānantyāyatana or Ākāsānañcāyatana "Sphere of Infinite Space". In this sphere formless beings dwell meditating upon space or extension (ākāśa) as infinitely pervasive. Total life span on this realm in human years - 20,000 Maha Kalpa. This realm is placed 5,580,000 Yodun above the Akanita Brahma Loka — Highest plane of pure abodes.

Form Realm (Rūpadhātu)

The Rūpadhātu or "Form realm" is, as the name implies, the first of the physical realms; its inhabitants all have a location and bodies of a sort, though those bodies are composed of a subtle substance which is of itself invisible to the inhabitants of the Kāmadhātu. According to the Janavasabha Sutta, when a brahma (a being from the Brahma-world of the Rūpadhātu) wishes to visit a deva of the Trāyastriṃśa heaven (in the Kāmadhātu), he has to assume a "grosser form" in order to be visible to them. There are 17-22 Rūpadhātu in Buddhism texts, the most common saying is 18.

The beings of the Form realm are not subject to the extremes of pleasure and pain, or governed by desires for things pleasing to the senses, as the beings of the Kāmadhātu are. The bodies of Form realm beings do not have sexual distinctions.

Like the beings of the Ārūpyadhātu, the dwellers in the Rūpadhātu have minds corresponding to the dhyānas (Pāli: jhānas). In their case it is the four lower dhyānas or rūpadhyānas. However, although the beings of the Rūpadhātu can be divided into four broad grades corresponding to these four dhyānas, each of them is subdivided into further grades, three for each of the four dhyānas and five for the Śuddhāvāsa devas, for a total of seventeen grades (the Theravāda tradition counts one less grade in the highest dhyāna for a total of sixteen).

Physically, the Rūpadhātu consists of a series of planes stacked on top of each other, each one in a series of steps half the size of the previous one as one descends. In part, this reflects the fact that the devas are also thought of as physically larger on the higher planes. The highest planes are also broader in extent than the ones lower down, as discussed in the section on *Sahasra cosmology*.

The height of these planes is expressed in *yojanas*, a measurement of very uncertain length, but sometimes taken to be about 4,000 times the height of a man, and so approximately 4.54 miles (7.31 km).

Pure Abodes

The Śuddhāvāsa (Pāli: Suddhāvāsa; Tib: *gnas gtsang ma*) worlds, or "Pure Abodes", are distinct from the other worlds of the Rūpadhātu in that they do not house beings who have been born there through ordinary merit or meditative attainments, but only those Anāgāmins ("Non-returners") who are already on the path to Arhat-hood and who will attain enlightenment directly from the Śuddhāvāsa worlds without being reborn in a lower plane. Every Śuddhāvāsa deva is therefore a protector of Buddhism. (Brahma Sahampati, who appealed to the newly enlightened Buddha to teach, was an Anagami under the previous Buddha). Because a Śuddhāvāsa deva will never be re-born outside the Śuddhāvāsa worlds, no Bodhisattva is ever born in these worlds, as a Bodhisattva must ultimately be reborn as a human being.

Since these devas rise from lower planes only due to the teaching of a Buddha, they can remain empty for very long periods if no Buddha arises. However, unlike the lower worlds, the Śuddhāvāsa worlds are never destroyed by natural catastrophe. The Śuddhāvāsa devas predict the coming of a Buddha and, taking the guise of Brahmins, reveal to human beings the signs by which a Buddha can be recognized. They also ensure that a Bodhisattva in his last life will see the four signs that will lead to his renunciation.

The five Śuddhāvāsa worlds are:

- Akanistha or Akanistha – World of devas "equal in rank" (literally: having no one as the youngest). The highest of all the Rūpadhātu worlds, it is often used to refer to the highest extreme of the universe. The current Śakra will eventually be born there. The duration of life in Akaniṣṭha is 16,000 kalpas (Vibhajyavāda tradition). Mahesvara the ruler of the three realms of samsara is said to dwell here. The height of this world is 167,772,160 yojanas above the Earth (approximately the distance of Saturn from Earth).

- Sudarśana or Sudassī – The "clear-seeing" devas live in a world similar to and friendly with the Akaniṣṭha world. The height of this world is 83,886,080 yojanas above the Earth (approximately the distance of Jupiter from Earth).

- Sudrśa or Sudassa – The world of the "beautiful" devas are said to be the place of re-birth for five kinds of anāgāmins. The height of this world is 41,943,040 yojanas above the Earth.

- Atapa or Atappa – The world of the "untroubled" devas, whose company those of lower realms wish for. The height of this world is 20,971,520 yojanas above the Earth (approximately the distance of Sun from Earth).

- Avrha or Aviha – The world of the "not falling" devas, perhaps the most common destination for reborn Anāgāmins. Many achieve arhatship directly in this world, but some pass away and are reborn in sequentially higher worlds of the Pure Abodes until they are at last reborn in the Akaniṣṭha world. These are called in Pāli uddhaṃsotas, "those whose stream goes upward". The duration of life in Avṛha is 1,000 kalpas (Vibhajyavāda tradition). The height of this world is 10,485,760 yojanas above the Earth (approximately the distanceof Mars from Earth).

Bṛhatphala Worlds

The mental state of the devas of the Bṛhatphala worlds corresponds to the fourth dhyāna, and is characterized by equanimity (upekṣā). The Bṛhatphala worlds form the upper limit to the destruction of the universe by wind at the end of a mahākalpa), that is, they are spared such destruction.

- Asaññasatta (Sanskrit: Asaṃjñasattva) (Vibhajyavāda tradition only) – "Unconscious beings", devas who have attained a high dhyāna (similar to that of the Formless Realm), and, wishing to avoid the perils of perception, have achieved a state of non-perception in which they endure for a time. After a while, however, perception arises again and they fall into a lower state.

- Brhatphala or Vehapphala (Tib: *'bras bu che*) – Devas "having great fruit". Their lifespan is 500 mahākalpas. (Vibhajyavāda tradition). Some Anāgāmins are reborn here. The height of this world is 5,242,880 yojanas above the Earth.(approximately the distance of Venus from Earth)

- Punyaprasava (Sarvāstivāda tradition only; Tib: *bsod nams skyes*) – The world of the devas who are the "offspring of merit". The height of this world is 2,621,440 yojanas above the Earth.

- Anabhraka (Sarvāstivāda tradition only; Tib: *sprin med*) – The world of the "cloudless" devas. The height of this world is 1,310,720 yojanas above the Earth.

Śubhakṛtsna Worlds

The mental state of the devas of the Śubhakṛtsna worlds corresponds to the third dhyāna, and is characterized by a quiet joy (sukha). These devas have bodies that radiate a steady light. The Śubhakṛtsna worlds form the upper limit to the destruction of the universe by water at the end of a mahākalpa, that is, the flood of water does not rise high enough to reach them.

- Śubhakṛtsna or Subhakinna / Subhakinha (Tib: *dge rgyas*) – The world of devas of "total beauty". Their lifespan is 64 mahākalpas (some sources: 4 mahākalpas) according to the Vibhajyavāda tradition. 64 mahākalpas is the interval between destructions of the universe by wind, including the Śubhakṛtsna worlds. The height of this world is 655,360 yojanas above the Earth.

- Apramāṇaśubha or Appamāṇasubha (Tib: *tshad med dge*) – The world of devas of "limitless beauty". Their lifespan is 32 mahākalpas (Vibhajyavāda tradition). They possess "faith, virtue, learning, munificence and wisdom". The height of this world is 327,680 yojanas above the Earth.

- Parīttaśubha or Parittasubha (Tib: *dge chung*) – The world of devas of "limited beauty". Their lifespan is 16 mahākalpas. The height of this world is 163,840 yojanas above the Earth.

Ābhāsvara Worlds

The mental state of the devas of the Ābhāsvara worlds corresponds to the second dhyāna, and is characterized by delight (prīti) as well as joy (sukha); the Ābhāsvara devas are said to shout aloud in their joy, crying *aho sukham!* ("Oh joy!"). These devas have bodies that emit flashing rays of light like lightning. They are said to have similar bodies (to each other) but diverse perceptions.

The Ābhāsvara worlds form the upper limit to the destruction of the universe by fire at the end of a mahākalpa, that is, the column of fire does not rise high enough to reach them. After the destruction of the world, at the beginning of the vivartakalpa, the worlds are first populated by beings reborn from the Ābhāsvara worlds.

- Ābhāsvara or Ābhassara (Tib: *'od gsal*) – The world of devas "possessing splendor". The lifespan of the Ābhāsvara devas is 8 mahākalpas (others: 2 mahākalpas). Eight mahākalpas is the interval between destructions of the universe by water, which includes the Ābhāsvara worlds. The height of this world is 81,920 yojanas above the Earth.

- Apramānābha or Appamānābha (Tib: *tshad med 'od*) – The world of devas of "limitless light", a concept on which they meditate. Their lifespan is 4 mahākalpas. The height of this world is 40,960 yojanas above the Earth.

- Parīttābha or Parittābha (Tib: *'od chung*) – The world of devas of "limited light". Their lifespan is 2 mahākalpas. The height of this world is 20,480 yojanas above the Earth.

Brahmā Worlds

The mental state of the devas of the Brahmā worlds corresponds to the first dhyāna, and is characterized by observation (vitarka) and reflection (vicāra) as well as delight (prīti) and joy (sukha). The Brahmā worlds, together with the other lower worlds of the universe, are destroyed by fire at the end of a mahākalpa.

- Mahābrahmā – the world of "Great Brahmā", believed by many to be the creator of the world, and having as his titles "Brahmā, Great Brahmā, the Conqueror, the Unconquered, the All-Seeing, All-Powerful, the Lord, the Maker and Creator, the Ruler, Appointer and Orderer, Father of All That Have Been and Shall Be." According to the Brahmajāla Sutta (DN.1), a Mahābrahmā is a being from the Ābhāsvara worlds who falls into a lower world through exhaustion of his merits and is reborn alone in the Brahma-world; forgetting his former existence, he imagines himself to have come into existence without cause. Note that even such a high-ranking deity has no intrinsic knowledge of the worlds above his own. Mahābrahmā is $1\frac{1}{2}$ yojanas tall. His lifespan variously said to be 1 kalpa (Vibhajyavāda tradition) or $1\frac{1}{2}$ kalpas long (Sarvāstivāda tradition), although it would seem that it could be no longer than $\frac{3}{4}$ of a mahākalpa, i.e., all of the mahākalpa except for the Saṃvartasthā-yikalpa, because that is the total length of time between the rebuilding of the lower world and its destruction. It is unclear what period of time "kalpa" refers to in this case. The height of this world is 10,240 yojanas above the Earth.

- Brahmapurohita (Tib: *tshangs 'khor*) – the "Ministers of Brahmā" are beings, also originally from the Ābhāsvara worlds, that are born as companions to Mahābrahmā after he has spent some time alone. Since they arise subsequent to his thought of a desire for companions, he believes himself to be their creator, and they likewise believe him to be their creator and lord. They are 1 yojana in height and their lifespan is variously said to be $\frac{1}{2}$ of a kalpa (Vibhajyavāda tradition) or a whole kalpa (Sarvāstivāda tradition). If they are later reborn in a lower world, and come to recall some part of their last existence, they teach the doctrine of Brahmā as creator as a revealed truth. The height of this world is 5,120 yojanas above the Earth.

- Brahmapārisadya or Brahmapārisajja (Tib: *tshangs ris*) – the "Councilors of Brahmā" or the devas "belonging to the assembly of Brahmā". They are also called Brahmakāyika, but this name can be used for any of the inhabitants of the Brahma-worlds. They are half a yojana in height and their lifespan is variously said to be $\frac{1}{3}$ of a kalpa (Vibhajyavāda tradition) or $\frac{1}{2}$ of a kalpa (Sarvāstivāda tradition). The height of this world is 2,560 yojanas above the Earth.

Desire Realm (Kāmadhātu)

The beings born in the Kāmadhātu differ in degree of happiness, but they are all, other than Anāgāmi, Arhat and Buddhas, under the domination of Māra and are bound by sensual desire, which causes them suffering.

Heavens

The following four worlds are bounded planes, each 80,000 yojanas square, which float in the air above the top of Mount Sumeru. Although all of the worlds inhabited by devas (that is, all the worlds down to the Cāturmahārājikakāyika world and sometimes including the Asuras) are sometimes called "heavens", in the western sense of the word the term best applies to the four worlds listed below:

- Parinirmita-vaśavartin or Paranimmita-vasavatti – The heaven of devas "with power over (others') creations". These devas do not create pleasing forms that they desire for themselves, but their desires are fulfilled by the acts of other devas who wish for their favor. The ruler of this world is called Vaśavartin (Pāli: Vasavatti), who has longer life, greater beauty, more power and happiness and more delightful sense-objects than the other devas of his world. This world is also the home of the devaputra (being of divine race) called Māra, who endeavors to keep all beings of the Kāmadhātu in the grip of sensual pleasures. Māra is also sometimes called Vaśavartin, but in general these two dwellers in this world are kept distinct. The beings of this world are 4,500 feet (1,400 m) tall and live for 9,216,000,000 years (Sarvāstivāda tradition). The height of this world is 1,280 yojanas above the Earth.

- Nirmānarati or Nimmānaratī– The world of devas "delighting in their creations". The devas of this world are capable of making any appearance to please themselves. The lord of this world is called Sunirmita (Pāli Sunimmita); his wife is the rebirth of Visākhā, formerly the chief of the upāsikās (female lay devotees) of the Buddha. The beings of this world are 3,750 feet (1,140 m) tall and live for 2,304,000,000 years (Sarvāstivāda tradition). The height of this world is 640 yojanas above the Earth.

- Tuṣita or Tusita – The world of the "joyful" devas. This world is best known for being the world in which a Bodhisattva lives before being reborn in the world of humans. Until a few thousand years ago, the Bodhisattva of this world was Śvetaketu (Pāli: Setaketu), who was reborn as Siddhārtha, who would become the Buddha Śākyamuni; since then the Bodhisattva has been Nātha (or Nāthadeva) who will be reborn as Ajita and will become the Buddha Maitreya (Pāli Metteyya). While this Bodhisattva is the foremost of the dwellers in Tuṣita, the ruler of this world is another deva called Santuṣita (Pāli: Santusita). The beings of this world are 3,000 feet (910 m) tall and live for 576,000,000 years (Sarvāstivāda tradition). The height of this world is 320 yojanas above the Earth.

- Yāma – Sometimes called the "heaven without fighting", because it is the lowest of the heavens to be physically separated from the tumults of the earthly world. These devas live in the air, free of all difficulties. Its ruler is the deva Suyāma; according to some, his wife is the rebirth of Sirimā, a courtesan of Rājagṛha in the Buddha's time who was generous to the monks. The beings of this world are 2,250 feet (690 m) tall and live for 144,000,000 years (Sarvāstivāda tradition). The height of this world is 160 yojanas above the Earth.

Worlds of Sumeru

The world-mountain of Sumeru is an immense, strangely shaped peak which arises in the center of the world, and around which the Sun and Moon revolve. Its base rests in a vast ocean, and it is surrounded by several rings of lesser mountain ranges and oceans. The three worlds listed below are all located on, or around, Sumeru: the Trāyastriṃśa devas live on its peak, the Cāturmahārājikakāyika devas live on its slopes, and the Asuras live in the ocean at its base. Sumeru and its surrounding oceans and mountains are the home not just of these deities, but also vast assemblies of beings of popular mythology who only rarely intrude on the human world.

- Trāyastrimśa or Tāvatimsa – The world "of the Thirty-three (devas)" is a wide flat space on the top of Mount Sumeru, filled with the gardens and palaces of the devas. Its ruler is Śakra devānām indra, "Śakra, lord of the devas". Besides the eponymous Thirty-three devas, many other devas and supernatural beings dwell here, including the attendants of the devas and many apsarases (nymphs). The beings of this world are 1,500 feet (460 m) tall and live for 36,000,000 years (Sarvāstivāda tradition) or 3/4 of a yojana tall and live for 30,000,000 years (Vibhajyavāda tradition). The height of this world is 80 yojanas above the Earth.

- Cāturmahārājikakāyika or Cātummahārājika – The world "of the Four Great Kings" is found on the lower slopes of Mount Sumeru, though some of its inhabitants live in the air around the mountain. Its rulers are the four Great Kings of the name, Virūḍhaka, Dhṛtarāṣṭra, Virūpākṣa, and their leader Vaiśravaṇa. The devas who guide the Sun and Moon are also considered part of this world, as are the retinues of the four kings, composed of Kumbhāṇḍas (dwarfs), Gandharvas (fairies), Nāgas (dragons) and Yakṣas (goblins). The beings of this world are 750 feet (230 m) tall and live for 9,000,000 years (Sarvāstivāda tradition) or 90,000 years (Vibhajyavāda tradition). The height of this world is from sea level up to 40 yojanas above the Earth.

- Asura – The world of the Asuras is the space at the foot of Mount Sumeru, much of which is a deep ocean. It is not the Asuras' original home, but the place they found themselves after they were hurled, drunken, from Trāyastriṃśa where they had formerly lived. The

Asuras are always fighting to regain their lost kingdom on the top of Mount Sumeru, but are unable to break the guard of the Four Great Kings. The Asuras are divided into many groups, and have no single ruler, but among their leaders are Vemacitrin (Pāli: Vepacitti) and Rāhu.

Earthly Realms

- Manuyaloka – This is the world of humans and human-like beings who live on the surface of the earth. The mountain-rings that engird Sumeru are surrounded by a vast ocean, which fills most of the world. The ocean is in turn surrounded by a circular mountain wall called Cakravāḍa (Pāli: Cakkavāḷa) which marks the horizontal limit of the world. In this ocean there are four continents which are, relatively speaking, small islands in it. Because of the immenseness of the ocean, they cannot be reached from each other by ordinary sailing vessels, although in the past, when the cakravartin kings ruled, communication between the continents was possible by means of the treasure called the cakraratna (Pāli cakkaratana), which a cakravartin and his retinue could use to fly through the air between the continents. The four continents are:

  ○ Jambudvīpa or Jambudīpa is located in the south and is the dwelling of ordinary human beings. It is said to be shaped "like a cart", or rather a blunt-nosed triangle with the point facing south. (This description probably echoes the shape of the coastline of southern India.) It is 10,000 yojanas in extent (Vibhajyavāda tradition) or has a perimeter of 6,000 yojanas (Sarvāstivāda tradition) to which can be added the southern coast of only 3 ½ yojanas' length. The continent takes its name from a giant Jambu tree (Syzygium cumini), 100 yojanas tall, which grows in the middle of the continent. Every continent has one of these giant trees. All Buddhas appear in Jambudvīpa. The people here are five to six feet tall and their length of life varies between 10 to power 140 years (Asankya Aayu) and 10 years.

  ○ Pūrvavideha or Pubbavideha is located in the east, and is shaped like a semicircle with the flat side pointing westward (i.e., towards Sumeru). It is 7,000 yojanas in extent (Vibhajyavāda tradition) or has a perimeter of 6,350 yojanas of which the flat side is 2,000 yojanas long (Sarvāstivāda tradition). Its tree is the acacia. The people here are about 12 feet (3.7 m) tall and they live for 250 years.

  ○ Aparagodānīya or Aparagoyāna is located in the west, and is shaped like a circle with a circumference of about 7,500 yojanas (Sarvāstivāda tradition). The tree of this continent is a giant Kadamba tree. The human inhabitants of this continent do not live in houses but sleep on the ground. They are about 24 feet (7.3 m) tall and they live for 500 years.

  ○ Uttarakuru is located in the north, and is shaped like a square. It has a perimeter of 8,000 yojanas, being 2,000 yojanas on each side. This continent's tree is called a kalpavṛkṣa (Pāli: kapparukkha) or kalpa-tree, because it lasts for the entire kalpa. The inhabitants of Uttarakuru have cities built in the air. They are said to be extraordinarily wealthy, not needing to labor for a living – as their food grows by itself – and having no private property. They are about 48 feet (15 m) tall and live for 1,000 years, and they are under the protection of Vaiśravaṇa.

- Tiryagyoni-loka or Tiracchāna-yoni – This world comprises all members of the animal kingdom that are capable of feeling suffering, regardless of size.

- Pretaloka or Petaloka – The pretas, or "hungry ghosts", are mostly dwellers on earth, though due to their mental state they perceive it very differently from humans. They live for the most part in deserts and wastelands.

Hells (Narakas)

Naraka or Niraya is the name given to one of the worlds of greatest suffering, usually translated into English as "hell" or "purgatory". As with the other realms, a being is born into one of these worlds as a result of his karma, and resides there for a finite length of time until his karma has achieved its full result, after which he will be reborn in one of the higher worlds as the result of an earlier karma that had not yet ripened. The mentality of a being in the hells corresponds to states of extreme fear and helpless anguish in humans.

Physically, Naraka is thought of as a series of layers extending below Jambudvīpa into the earth. There are several schemes for counting these Narakas and enumerating their torments. One of the more common is that of the Eight Cold Narakas and Eight Hot Narakas.

Cold Narakas

- Arbuda – the "blister" Naraka

- Nirarbuda – the "burst blister" Naraka

- Atata – the Naraka of shivering

- Hahava – the Naraka of lamentation

- Huhuva – the Naraka of chattering teeth

Each lifetime in these Narakas is twenty times the length of the one before it.

Hot Narakas

- Sañjīva – the "reviving" Naraka. Life in this Naraka is $162*10^{10}$ years long.

- Kālasūtra – the "black thread" Naraka. Life in this Naraka is $1296*10^{10}$ years long.

- Samghāta – the "crushing" Naraka. Life in this Naraka is $10,368*10^{10}$ years long.

- Raurava – the "screaming" Naraka. Life in this Naraka is $82,944*10^{10}$ years long.

- Mahāraurava – the "great screaming" Naraka. Life in this Naraka is $663,552*10^{10}$ years long.

- Tapana – the "heating" Naraka. Life in this Naraka is $5,308,416*10^{10}$ years long.

- Pratāpana – the "great heating" Naraka. Life in this Naraka is $42,467,328*10^{10}$ years long.

- Avīci – the "uninterrupted" Naraka. Life in this Naraka is $339,738,624*10^{10}$ years long.

The Foundations of the Earth

All of the structures of the earth, Sumeru and the rest, extend downward to a depth of 80,000 yojanas below sea level – the same as the height of Sumeru above sea level. Below this is a layer of "golden earth", a substance compact and firm enough to support the weight of Sumeru. It is 320,000 yojanas in depth and so extends to 400,000 yojanas below sea level. The layer of golden earth in turn rests upon a layer of water, which is 8,000,000 yojanas in depth, going down to 8,400,000 yojanas below sea level. Below the layer of water is a "circle of wind", which is 16,000,000 yojanas in depth and also much broader in extent, supporting 1,000 different worlds upon it. Yojanas are equivalent to about 13 km (8 mi).

Sahasra Cosmology

Sahasra means "one thousand". All of the planes, from the plane of neither perception nor non-perception (nevasanna-asanna-ayatana) down to the Avici – the "uninterrupted" or "unceasing" (avici literally means "without interval") niraya – constitutes the single world-system, cakkavala (intimating something circular, a "wheel", but the etymology is uncertain), described above. In modern parlance it would be called a 'universe', or 'solar system'.

A collection of one thousand solar systems are called a "thousandfold minor world-system" (culanika lokhadhatu). Or small chiliocosm.

A collection of 1,000 times 1,000 world-systems (one thousand squared) is a "thousandfold to the second power middling world-system" (dvisahassi majjhima lokadhatu). Or medium dichiliocosm.

The largest grouping, which consists of one thousand cubed world-systems, is called the "tisahassi mahasassi lokadhatu".

The Tathagata, if he so wished, could effect his voice throughout a great trichiliocosm. He does so by suffusing the trichiliocosm with his radiance, which at the point the inhabitants of those world-system will perceive this light, and then proceeds to extend his voice throughout that realm.

Maha Kalpa

The word kalpa, means 'moment'. A maha kalpa consists of four moments (kalpa), the first of which is creation. The creation moment consists of the creation of the "receptacle", and the descent of beings from higher realms into more coarse forms of existence. During the rest of the creation moment, the world is populated. Human beings who exist at this point have no limit on their lifespan. The second moment is the duration moment, the start of this moment is signified by the first sentient being to enter hell (niraya), the hells and nirayas not existing or being empty prior to this moment. The duration moment consists of twenty "intermediate" moments (antarakappas), which unfold in a drama of the human lifespan descending from 80,000 years to 10, and then back up to 80,000 again. The interval between 2 of these "intermediate" moments is the "seven day purge", in which a variety of humans will kill each other (not knowing or recognizing each other), some humans will go into hiding. At the end of this purge, they will emerge from hiding and repopulate the world. As of May 2015, it seems the lifespan of humans is 80 years, during the time of Gotama Buddha it was 100 years. After this purge, the lifespan will increase to 80,000, reach its peak and descend, at which point the purge will happen again.

Within the duration 'moment', this purge and repeat cycle seems to happen around 18 times, the first "intermediate" moment consisting only of the descent from 80,000—the second intermediate moment consisting of a rise and descent, and the last consisting only of an ascent.

After the duration 'moment' is the dissolution moment, the hells will gradually be emptied, as well as all coarser forms of existence. The beings will flock to the form realms (rupa dhatu), a destruction of fire occurs, sparing everything from the realms of the 'radiant' gods and above (abha deva).

After 7 of these destructions by 'fire', a destruction by water occurs, and everything from the realms of the 'pleasant' gods and above is spared (subha deva).

After 64 of these destructions by fire and water, that is—56 destructions by fire, and 7 by water—a destruction by wind occurs, this eliminates everything below the realms of the 'fruitful' devas (vehapphala devas, literally of "great fruit"). The pure abodes (suddhavasa, meaning something like pure, unmixed, similar to the connotation of "pure bred German shepherd"), are never destroyed. Although without the appearance of a Buddha, these realms may remain empty for a long time. It should be noted that the inhabitants of these realms have exceedingly long life spans.

The formless realms are never destroyed because they do not consist of form (rupa). The reason the world is destroyed by fire, water and wind, and not earth is because earth is the 'receptacle'.

After the dissolution moment, this particular world system remains dissolved for a long time, this is called the 'empty' moment, but the more accurate term would be "the state of being dissolved". The beings that inhabited this realm formerly will migrate to other world systems, and perhaps return if their journeys lead here again.

Temporal Cosmology

Buddhist temporal cosmology describes how the universe comes into being and is dissolved. Like other Indian cosmologies, it assumes an infinite span of time and is cyclical. This does not mean that the same events occur in identical form with each cycle, but merely that, as with the cycles of day and night or summer and winter, certain natural events occur over and over to give some structure to time.

The basic unit of time measurement is the mahākalpa or "Great Eon". The length of this time in human years is never defined exactly, but it is meant to be very long, to be measured in billions of years if not longer.

A mahākalpa is divided into four kalpas or "eons" , each distinguished from the others by the stage of evolution of the universe during that kalpa. The four kalpas are:

- Vivartakalpa "Eon of evolution" – during this kalpa the universe comes into existence.

- Vivartasthāyikalpa "Eon of evolution-duration" – during this kalpa the universe remains in existence in a steady state.

- Samvartakalpa "Eon of dissolution" – during this kalpa the universe dissolves.

- Samvartasthāyikalpa "Eon of dissolution-duration" – during this kalpa the universe remains in a state of emptiness.

Each one of these kalpas is divided into twenty antarakalpas each of about the same length. For the Saṃvartasthāyikalpa this division is merely nominal, as nothing changes from one antarakalpa to the next; but for the other three kalpas it marks an interior cycle within the kalpa.

Vivartakalpa

The Vivartakalpa begins with the arising of the primordial wind, which begins the process of building up the structures of the universe that had been destroyed at the end of the last mahākalpa. As the extent of the destruction can vary, the nature of this evolution can vary as well, but it always takes the form of beings from a higher world being born into a lower world. The example of a Mahābrahmā being the rebirth of a deceased Ābhāsvara deva is just one instance of this, which continues throughout the Vivartakalpa until all the worlds are filled from the Brahmaloka down to Naraka. During the Vivartakalpa the first humans appear; they are not like present-day humans, but are beings shining in their own light, capable of moving through the air without mechanical aid, living for a very long time, and not requiring sustenance; they are more like a type of lower deity than present-day humans are.

Over time, they acquire a taste for physical nutriment, and as they consume it, their bodies become heavier and more like human bodies; they lose their ability to shine, and begin to acquire differences in their appearance, and their length of life decreases. They differentiate into two sexes and begin to become sexually active. Then greed, theft and violence arise among them, and they establish social distinctions and government and elect a king to rule them, called Mahāsammata, "the great appointed one". Some of them begin to hunt and eat the flesh of animals, which have by now come into existence.

Vivartasthāyikalpa

First Antarakalpa

The Vivartasthāyikalpa begins when the first being is born into Naraka, thus filling the entire universe with beings. During the first antarakalpa of this eon, human lives are declining from a vast but unspecified number of years (but at least several tens of thousands of years) toward the modern lifespan of less than 100 years. At the beginning of the antarakalpa, people are still generally happy. They live under the rule of a universal monarch or "wheel-turning king", who conquer. The Mahāsudassana-sutta (DN.17) tells of the life of a cakravartin king, Mahāsudassana (Sanskrit: Mahāsudarśana) who lived for 336,000 years. The Cakkavatti-sīhanāda-sutta (DN.26) tells of a later dynasty of cakravartins, Daḷhanemi (Sanskrit: Dṛḍhanemi) and five of his descendants, who had a lifespan of over 80,000 years. The seventh of this line of cakravartins broke with the traditions of his forefathers, refusing to abdicate his position at a certain age, pass the throne on to his son, and enter the life of a śramaṇa. As a result of his subsequent misrule, poverty increased; as a result of poverty, theft began; as a result of theft, capital punishment was instituted; and as a result of this contempt for life, murders and other crimes became rampant.

The human lifespan now quickly decreased from 80,000 to 100 years, apparently decreasing by about half with each generation (this is perhaps not to be taken literally), while with each generation other crimes and evils increased: lying, greed, hatred, sexual misconduct, disrespect for elders. During this period, according to the Mahāpadāna-sutta (DN.14) three of the four Buddhas of this antarakalpa lived: Krakucchanda Buddha (Pāli: Kakusandha), at the time when the lifespan was 40,000 years; Kanakamuni Buddha (Pāli: Konāgamana) when the lifespan was 30,000 years; and Kāśyapa Buddha (Pāli: Kassapa) when the lifespan was 20,000 years.

Our present time is taken to be toward the end of the first antarakalpa of this Vivartasthāyikalpa, when the lifespan is less than 100 years, after the life of Śākyamuni Buddha (Pāli: Sakyamuni), who lived to the age of 80.

The remainder of the antarakalpa is prophesied to be miserable: lifespans will continue to decrease, and all the evil tendencies of the past will reach their ultimate in destructiveness. People will live no longer than ten years, and will marry at five; foods will be poor and tasteless; no form of morality will be acknowledged. The most contemptuous and hateful people will become the rulers. Incest will be rampant. Hatred between people, even members of the same family, will grow until people think of each other as hunters do of their prey.

Eventually a great war will ensue, in which the most hostile and aggressive will arm themselves and go out to kill each other. The less aggressive will hide in forests and other secret places while the war rages. This war marks the end of the first antarakalpa.

Second Antarakalpa

At the end of the war, the survivors will emerge from their hiding places and repent their evil habits. As they begin to do good, their lifespan increases, and the health and welfare of the human race will also increase with it. After a long time, the descendants of those with a 10-year lifespan will live for 80,000 years, and at that time there will be a cakravartin king named Saṅkha. During his reign, the current bodhisattva in the Tuṣita heaven will descend and be reborn under the name of Ajita. He will enter the life of a śramaṇa and will gain perfect enlightenment as a Buddha; and he will then be known by the name of Maitreya (Pāli: Metteyya).

After Maitreya's time, the world will again worsen, and the lifespan will gradually decrease from 80,000 years to 10 years again, each antarakalpa being separated from the next by devastating war, with peaks of high civilization and morality in the middle. After the 19th antarakalpa, the lifespan will increase to 80,000 and then not decrease, because the Vivartasthāyikalpa will have come to an end.

Saṃvartakalpa

The Saṃvartakalpa begins when beings cease to be born in Naraka. This cessation of birth then proceeds in reverse order up the vertical cosmology, i.e., pretas then cease to be born, then animals, then humans, and so on up to the realms of the deities.

When these worlds as far as the Brahmaloka are devoid of inhabitants, a great fire consumes the entire physical structure of the world. It burns all the worlds below the Ābhāsvara worlds. When they are destroyed, the Saṃvartasthāyikalpa begins.

Saṃvartasthāyikalpa

There is nothing to say about the Saṃvartasthāyikalpa, since nothing happens in it below the Ābhāsvara worlds. It ends when the primordial wind begins to blow and build the structure of the worlds up again.

Other Destructions

The destruction by fire is the normal type of destruction that occurs at the end of the Saṃvartakalpa. But every eighth mahākalpa, after seven destructions by fire, there is a destruction by water. This is more devastating, as it eliminates not just the Brahma worlds but also the Ābhāsvara worlds.

Every sixty-fourth mahākalpa, after fifty six destructions by fire and seven destructions by water, there is a destruction by wind. This is the most devastating of all, as it also destroys the Śubhakṛtsna worlds. The higher worlds are never destroyed.

Mahayana Views

Mahayana Buddhism accepts the cosmology as above. A cosmology with some difference is further explained in Chapter 5 of the Avatamsaka Sutra.

Creationism

Creationism is the religious belief that the universe and life originated "from specific acts of divine creation," as opposed to the scientific conclusion that they came about through natural processes. The first use of the term "creationist" to describe a proponent of creationism is found in an 1856 letter of Charles Darwin describing those who objected on religious grounds to the emerging science of evolution. Creationism covers a spectrum of views including *evolutionary creationism*, a theological variant of theistic evolution which asserts that both evolutionary science and a belief in creation are true, but the term is commonly used for literal creationists who reject various aspects of science, and instead promote belief in pseudoscience.

Literal creationists base their beliefs on a fundamentalist reading of religious texts, including the creation myths found in Genesis and the Quran. For young Earth creationists, these beliefs are based on a literalist interpretation of the Genesis creation narrative and rejection of the scientific theory of evolution. Literalist creationists believe that evolution cannot adequately account for the history, diversity, and complexity of life on Earth. Pseudoscientific branches of creationism include creation science, flood geology, and intelligent design, as well as subsets of pseudoarchaeology, pseudohistory, and pseudolinguistics.

Biblical Basis

The basis for many creationists' beliefs is a literal or quasi-literal interpretation of the Old Testament, especially from stories from the book of Genesis:

- The Genesis creation narrative (Genesis 1–2) describes how God brings the Universe into being in a series of creative acts over six days and places the first man and woman (Adam and Eve) in a divine garden (the Garden of Eden). This story is the basis of Creationist cosmology and biology.

- The Genesis flood narrative (Genesis 6–9) tells how God destroys the world and all life through a great flood, saving representatives of each form of life by means of Noah's ark. This forms the basis of Creationist geology, better known as flood geology.

A further important element is the interpretation of the Biblical chronology, the elaborate system of life-spans, "generations," and other means by which the Bible measures the passage of events from the Creation (Genesis 1:1) to the Book of Daniel, the last biblical book in which it appears. Recent decades have seen attempts to de-link Creationism from the Bible and recast it as science: these include creation science and intelligent design There are also non-Christian forms of Creationism, notably Islamic Creationism and Hindu Creationism.

Types of Creationism

Several attempts have been made to categorize the different types of creationism, and create a "taxonomy" of creationists. Creationism (broadly construed) covers a spectrum of beliefs which have been categorized into the general types listed below.

Comparison of major creationist views					
	Acceptance in the US	Humanity	Biological species	Earth	Age of Universe
Young Earth creationism	40%	Directly created by God.	Directly created by God. Macroevolution does not occur.	Less than 10,000 years old. Re-shaped by global flood.	Less than 10,000 years old, but some hold this view only for our Solar System.
Gap creationism				Scientifically accepted age. Re-shaped by global flood.	Scientifically accepted age.
Progressive creationism		Directly created by God, based on primate anatomy.	Direct creation + evolution. No single common ancestor.	Scientifically accepted age. No global flood.	Scientifically accepted age.
Intelligent design	38%	Proponents hold various beliefs. (For example, Michael Behe accepts evolution from primates.)	Divine intervention at some point in the past, as evidenced by what intelligent-design creationists call "irreducible complexity."	Some adherents accept common descent, others not. Some claim the existence of Earth is the result of divine intervention.	Scientifically accepted age.
Theistic evolution (evolutionary creationism)		Evolution from primates.	Evolution from single common ancestor.	Scientifically accepted age. No global flood.	Scientifically accepted age.

Young Earth Creationism

Young Earth creationists believe that God created the Earth within the last ten thousand years, literally as described in the Genesis creation narrative, within the approximate time-frame of biblical genealogies (detailed for example in the Ussher chronology). Most young Earth creationists believe that the universe has a similar age as the Earth. A few assign a much older age to the universe than to Earth. Creationist cosmologies give the universe an age consistent with the Ussher chronology and other young Earth time frames. Other young Earth creationists believe that the Earth and the universe were created with the appearance of age, so that the world appears to be much older than it is, and that this appearance is what gives the geological findings and other methods of dating the Earth and the universe their much longer timelines.

The Institute for Creation Research (ICR) is a young-Earth creationist organization.

The Christian organizations Institute for Creation Research (ICR) and the Creation Research Society (CRS) both promote young Earth creationism in the US. Another organization with similar views, Answers in Genesis (AiG)—based in both the US and the United Kingdom—has opened the Creation Museum in Petersburg, Kentucky, to promote young Earth creationism. Creation Ministries International promotes young Earth views in Australia, Canada, South Africa, New Zealand, the US, and the UK. Among Roman Catholics, the Kolbe Center for the Study of Creation promotes similar ideas.

Old Earth Creationism

Old Earth creationism holds that the physical universe was created by God, but that the creation event described in the Book of Genesis is to be taken figuratively. This group generally believes that the age of the universe and the age of the Earth are as described by astronomers and geologists, but that details of modern evolutionary theory are questionable.

Old Earth creationism itself comes in at least three types:

Gap Creationism

Gap creationism, also called "restoration creationism," holds that life was recently created on a pre-existing old Earth. This version of creationism relies on a particular interpretation of Genesis 1:1–2. It is considered that the words *formless* and *void* in fact denote waste and ruin, taking into account the original Hebrew and other places these words are used in the Old Testament. Genesis 1:1–2 is consequently translated:

> "In the beginning God created the heaven and the earth." (Original act of creation.)

> "And the earth was without form, and void; and darkness was upon the face of the deep. And the Spirit of God moved upon the face of the waters."

Thus, the six days of creation (verse 3 onwards) start sometime after the Earth was "without form and void." This allows an indefinite "gap" of time to be inserted after the original creation of the universe, but prior to the creation according to Genesis, (when present biological species and humanity were created). Gap theorists can therefore agree with the scientific consensus regarding the age of the Earth and universe, while maintaining a literal interpretation of the biblical text.

Some gap creationists expand the basic version of creationism by proposing a "primordial creation" of biological life within the "gap" of time. This is thought to be "the world that then was" mentioned in 2 Peter 3:3–7. Discoveries of fossils and archaeological ruins older than 10,000 years are generally ascribed to this "world that then was," which may also be associated with Lucifer's rebellion. These views became popular with publications of Hebrew Lexicons such as *Strong's Concordance*, and Bible commentaries such as the *Scofield Reference Bible* and *The Companion Bible*.

Day-age Creationism

Day-age creationism states that the "six days" of the Book of Genesis are not ordinary 24-hour days, but rather much longer periods (for instance, each "day" could be the equivalent of millions, or billions of years of human time). Physicist Gerald Schroeder is one such proponent of this view. This version of creationism often states that the Hebrew word "yôm," in the context of Genesis 1, can be properly interpreted as "age." Some adherents claim we are still living in the seventh age ("seventh day").

Strictly speaking, day-age creationism is not so much a version of creationism as a hermeneutic option which may be combined with other versions of creationism such as progressive creationism.

Progressive Creationism

Progressive creationism holds that species have changed or evolved in a process continuously guided by God, with various ideas as to how the process operated—though it is generally taken that God directly intervened in the natural order at key moments in Earth history. This view accepts most of modern physical science including the age of the Earth, but rejects much of modern evolutionary biology or looks to it for evidence that evolution by natural selection alone is incorrect. Organizations such as Reasons To Believe, founded by Hugh Ross, promote this version of creationism.

Progressive creationism can be held in conjunction with hermeneutic approaches to the Genesis creation narrative such as the day-age creationism or framework/metaphoric/poetic views.

Philosophic and Scientific Creationism

Creation Science

Creation science, or initially scientific creationism, is a pseudoscience that emerged in the 1960s with proponents aiming to have young Earth creationist beliefs taught in school science classes as a counter to teaching of evolution. Common features of Creation science argument include: creationist cosmologies which accommodate a universe on the order of thousands of years old, criticism of radiometric dating through a technical argument about radiohalos, explanations for the fossil record as a record of the Genesis flood narrative, and explanations for the present diversity as a result of pre-designed genetic variability and partially due to the rapid degradation of the perfect genomes God placed in "created kinds" or "Baramin" due to mutations.

Neo-creationism

Neo-Creationists intentionally distance themselves from other forms of creationism, preferring to be known as wholly separate from creationism as a philosophy. Neo-creationism aims to restate creationism in terms more likely to be well received by the public, policy makers, educators and the scientific community. It aims to re-frame the debate over the origins of life in non-religious terms and without appeals to scripture, and to bring the debate before the public.

Neo-creationism sees ostensibly objective mainstream science as a dogmatically atheistic religion. Neo-creationists argue that the scientific method excludes certain explanations of phenomena, particularly where they point towards supernatural elements. They argue that this effectively excludes any possible religious insight from contributing to a scientific understanding of the universe. Neo-creationists also argue that science, as an "atheistic enterprise," lies at the root of many of contemporary society's ills including social unrest and family breakdown.

The intelligent design movement arguably represents the most recognized form of neo-creationism in the US. Unlike their philosophical forebears, neo-creationists largely do not believe in many of the traditional cornerstones of creationism such as a young Earth, or in a dogmatically literal interpretation of the Bible. Common to all forms of neo-creationism is a rejection of naturalism, usually made together with a tacit admission of supernaturalism, and an open and often hostile opposition to what they term "Darwinism," meaning evolution.

Intelligent Design

Intelligent design (ID) is the pseudoscientific view that "certain features of the universe and of living things are best explained by an intelligent cause, not an undirected process such as natural selection." All of its leading proponents are associated with the Discovery Institute, a think tank whose Wedge strategy aims to replace the scientific method with "a science consonant with Christian and theistic convictions" which accepts supernatural explanations. It is widely accepted in the scientific and academic communities that intelligent design is a form of creationism, and is sometimes referred to as "intelligent design creationism."

ID originated as a re-branding of creation science in an attempt to avoid a series of court decisions ruling out the teaching of creationism in American public schools, and the Discovery Institute has run a series of campaigns to change school curricula. In Australia, where curricula are under the control of state governments rather than local school boards, there was a public outcry when the notion of ID being taught in science classes was raised by the Federal Education Minister Brendan Nelson; the minister quickly conceded that the correct forum for ID, if it were to be taught, is in religious or philosophy classes.

In the US, teaching of intelligent design in public schools has been decisively ruled by a federal district court to be in violation of the Establishment Clause of the First Amendment to the United States Constitution. In Kitzmiller v. Dover, the court found that intelligent design is not science and "cannot uncouple itself from its creationist, and thus religious, antecedents," and hence cannot be taught as an alternative to evolution in public school science classrooms under the jurisdiction of that court. This sets a persuasive precedent, based on previous US Supreme Court decisions in *Edwards v. Aguillard* and *Epperson v. Arkansas* (1968), and by the application of the Lemon test, that creates a legal hurdle to teaching intelligent design in public school districts in other federal court jurisdictions.

Obscure and Largely Discounted Beliefs

In astronomy, the geocentric model (also known as geocentrism, or the Ptolemaic system), is a description of the Cosmos where Earth is at the orbital center of all celestial bodies. This model served as the predominant cosmological system in many ancient civilizations such as ancient Greece. As such, they assumed that the Sun, Moon, stars, and naked eye planets circled Earth, including the noteworthy systems of Aristotle and Ptolemy.

Articles arguing that geocentrism was the biblical perspective appeared in some early creation science newsletters associated with the Creation Research Society pointing to some passages in the Bible, which, when taken literally, indicate that the daily apparent motions of the Sun and the Moon are due to their actual motions around the Earth rather than due to the rotation of the Earth about its axis for example, Joshua 10:12 where the Sun and Moon are said to stop in the sky, and Psalms 93:1 where the world is described as immobile. Contemporary advocates for such religious beliefs include Robert Sungenis, co-author of the self-published *Galileo Was Wrong: The Church Was Right* (2006). These people subscribe to the view that a plain reading of the Bible contains an accurate account of the manner in which the universe was created and requires a geocentric worldview. Most contemporary creationist organizations reject such perspectives.

The Omphalos hypothesis argues that in order for the world to be functional, God must have created a mature Earth with mountains and canyons, rock strata, trees with growth rings, and so on; therefore *no* evidence that we can see of the presumed age of the Earth and age of the universe can be taken as reliable. The idea has seen some revival in the 20th century by some modern creationists, who have extended the argument to address the "starlight problem". The idea has been criticised as Last Thursdayism, and on the grounds that it requires a deliberately deceptive creator.

Theistic Evolution

Theistic evolution, or evolutionary creation, is a belief that "the personal God of the Bible created the universe and life through evolutionary processes." According to the American Scientific Affiliation:

A theory of theistic evolution (TE) – also called evolutionary creation – proposes that God's method of creation was to cleverly design a universe in which everything would naturally evolve. Usually the "evolution" in "theistic evolution" means Total Evolution – astronomical evolution (to form galaxies, solar systems,...) and geological evolution (to form the earth's geology) plus chemical evolution (to form the first life) and biological evolution (for the development of life) – but it can refer only to biological evolution.

Through the 19th century the term *creationism* most commonly referred to direct creation of individual souls, in contrast to traducianism. Following the publication of *Vestiges of the Natural History of Creation*, there was interest in ideas of Creation by divine law. In particular, the liberal theologian Baden Powell argued that this illustrated the Creator's power better than the idea of miraculous creation, which he thought ridiculous. When *On the Origin of Species* was published, the cleric Charles Kingsley wrote of evolution as "just as noble a conception of Deity." Darwin's view at the time was of God creating life through the laws of nature, and the book makes several references to "creation," though he later regretted using the term rather than calling it an unknown process. In America, Asa Gray argued that evolution is the secondary effect, or *modus operandi*, of the first cause, design, and published a pamphlet defending the book in theistic terms, *Natural Selection not inconsistent with Natural Theology*. Theistic evolution, also called, evolutionary creation, became a popular compromise, and St. George Jackson Mivart was among those accepting evolution but attacking Darwin's naturalistic mechanism. Eventually it was realised that supernatural intervention could not be a scientific explanation, and naturalistic mechanisms such as neo-Lamarckism were favoured as being more compatible with purpose than natural selection.

Some theists took the general view that, instead of faith being in opposition to biological evolution, some or all classical religious teachings about Christian God and creation are compatible with some or all of modern scientific theory, including specifically evolution; it is also known as "evolutionary creation." In Evolution versus Creationism, Eugenie Scott and Niles Eldredge state that it is in fact a type of evolution.

It generally views evolution as a tool used by God, who is both the first cause and immanent sustainer/upholder of the universe; it is therefore well accepted by people of strong theistic (as opposed to deistic) convictions. Theistic evolution can synthesize with the day-age creationist interpretation of the Genesis creation narrative; however most adherents consider that the first chapters of the Book of Genesis should not be interpreted as a "literal" description, but rather as a literary framework or allegory.

From a theistic viewpoint, the underlying laws of nature were designed by God for a purpose, and are so self-sufficient that the complexity of the entire physical universe evolved from fundamental particles in processes such as stellar evolution, life forms developed in biological evolution, and in the same way the origin of life by natural causes has resulted from these laws.

In one form or another, theistic evolution is the view of creation taught at the majority of mainline Protestant seminaries. For Roman Catholics, human evolution is not a matter of religious teaching, and must stand or fall on its own scientific merits. Evolution and the Roman Catholic Church are not in conflict. The Catechism of the Catholic Church comments positively on the theory of evolution, which is neither precluded nor required by the sources of faith, stating that scientific studies "have splendidly enriched our knowledge of the age and dimensions of the cosmos, the development of life-forms and the appearance of man." Roman Catholic schools teach evolution without controversy on the basis that scientific knowledge does not extend beyond the physical, and scientific truth and religious truth cannot be in conflict. Theistic evolution can be described as "creationism" in holding that divine intervention brought about the origin of life or that divine laws govern formation of species, though many creationists (in the strict sense) would deny that the position is creationism at all. In the creation–evolution_controversy controversy its proponents generally take the "evolutionist" side. This sentiment was expressed by Fr. George Coyne, (the Vatican's chief astronomer between 1978 and 2006):

...in America, creationism has come to mean some fundamentalistic, literal, scientific interpretation of Genesis. Judaic-Christian faith is radically creationist, but in a totally different sense. It is rooted in a belief that everything depends upon God, or better, all is a gift from God.

While supporting the methodological naturalism inherent in modern science, the proponents of theistic evolution reject the implication taken by some atheists that this gives credence to ontological materialism. In fact, many modern philosophers of science, including atheists, refer to the long-standing convention in the scientific method that observable events in nature should be explained by natural causes, with the distinction that it does not assume the actual existence or non-existence of the supernatural.

Religious Views

Christianity

As of 2006, most Christians around the world accepted evolution as the most likely explanation for the origins of species, and did not take a literal view of the Genesis creation myth. The United States is an exception where belief in religious fundamentalism is much more likely to affect attitudes towards evolution than it is for believers elsewhere. Political partisanship affecting religious belief may be a factor because political partisanship in the US is highly correlated with fundamentalist thinking, unlike in Europe.

Most contemporary Christian leaders and scholars from mainstream churches, such as Anglicans and Lutherans, consider that there is no conflict between the spiritual meaning of creation and the science of evolution. According to the former Archbishop of Canterbury, Rowan Williams, "...for most of the history of Christianity, and I think this is fair enough, most of the history of the Christianity there's been an awareness that a belief that everything depends on the creative act of God, is quite compatible with a degree of uncertainty or latitude about how precisely that unfolds in creative time."

Leaders of the Anglican and Roman Catholic churches have made statements in favor of evolutionary theory, as have scholars such as the physicist John Polkinghorne, who argues that evolution is one of the principles through which God created living beings. Earlier supporters of evolutionary theory include Frederick Temple, Asa Gray and Charles Kingsley who were enthusiastic supporters of Darwin's theories upon their publication, and the French Jesuit priest and geologist Pierre Teilhard de Chardin saw evolution as confirmation of his Christian beliefs, despite condemnation from Church authorities for his more speculative theories. Another example is that of Liberal theology, not providing any creation models, but instead focusing on the symbolism in beliefs of the time of authoring Genesis and the cultural environment.

Many Christians and Jews had been considering the idea of the creation history as an allegory (instead of historical) long before the development of Darwin's theory of evolution. For example, Philo, whose works were taken up by early Church writers, wrote that it would be a mistake to think that creation happened in six days, or in any set amount of time. Augustine of the late fourth century who was also a former neoplatonist argued that everything in the universe was created by God at the same moment in time (and not in six days as a literal reading of the Book of Genesis would seem to require); It appears that both Philo and Augustine felt uncomfortable with the idea of a seven-day creation because it detracted from the notion of God's omnipotence. In 1950, Pope Pius XII stated limited support for the idea in his encyclical *Humani generis*. In 1996, Pope John Paul II stated that "new knowledge has led to the recognition of the theory of evolution as more than a hypothesis," but, referring to previous papal writings, he concluded that "if the human body takes its origin from pre-existent living matter, the spiritual soul is immediately created by God."

In the US, Evangelical Christians have continued to believe in a literal Genesis. Members of evangelical Protestant (70%), Mormon (76%) and Jehovah's Witnesses (90%) denominations are the most likely to reject the evolutionary interpretation of the origins of life. However, the official website of the Jehovah's Witnesses states that while Jehovah's Witnesses believe that God created everything, they do not accept creationism, believing that some creationist views conflict with the Bible.

The historic Christian literal interpretation of creation requires the harmonization of the two creation stories, Genesis 1:1–2:3 and Genesis 2:4–25, for there to be a consistent interpretation. They sometimes seek to ensure that their belief is taught in science classes, mainly in American schools. Opponents reject the claim that the literalistic biblical view meets the criteria required to be considered scientific. Many religious groups teach that God created the Cosmos. From the days of the early Christian Church Fathers there were allegorical interpretations of the Book of Genesis as well as literal aspects.

Christian Science, a system of thought and practice derived from the writings of Mary Baker Eddy, interprets the Book of Genesis figuratively rather than literally. It holds that the material world is an illusion, and consequently not created by God: the only real creation is the spiritual realm, of which the material world is a distorted version. Christian Scientists regard the story of the creation in the Book of Genesis as having symbolic rather than literal meaning. According to Christian Science, both creationism and evolution are false from an absolute or "spiritual" point of view, as they both proceed from a (false) belief in the reality of a material universe. However, Christian Scientists do not oppose the teaching of evolution in schools, nor do they demand that alternative accounts be taught: they believe that both material science and literalist theology are concerned with the illusory, mortal and material, rather than the real, immortal and spiritual. With regard to material theories of creation, Eddy showed a preference for Darwin's theory of evolution over others.

Hinduism

According to Hindu creationism all species on Earth including humans have "devolved" or come down from a high state of pure consciousness. Hindu creationists claim that species of plants and animals are material forms adopted by pure consciousness which live an endless cycle of births and rebirths. Ronald Numbers says that: "Hindu Creationists have insisted on the antiquity of humans, who they believe appeared fully formed as long, perhaps, as trillions of years ago." Hindu creationism is a form of old Earth creationism, according to Hindu creationists the universe may even be older than billions of years. These views are based on the Vedas, the creation myths of which depict an extreme antiquity of the universe and history of the Earth.

Islam

Islamic creationism is the belief that the universe (including humanity) was directly created by God as explained in the Qur'an. It usually views the Book of Genesis as a corrupted version of God's message. The creation myths in the Qur'an are vaguer and allow for a wider range of interpretations similar to those in other Abrahamic religions.

Islam also has its own school of theistic evolutionism, which holds that mainstream scientific analysis of the origin of the universe is supported by the Qur'an. Some Muslims believe in evolutionary creation, especially among liberal movements within Islam.

Writing for *The Boston Globe*, Drake Bennett noted: "Without a Book of Genesis to account for ... Muslim creationists have little interest in proving that the age of the Earth is measured in the thousands rather than the billions of years, nor do they show much interest in the problem of the dinosaurs. And the idea that animals might evolve into other animals also tends to be less controversial, in part because there are passages of the Koran that seem to support it. But the issue of whether human beings are the product of evolution is just as fraught among Muslims." However, some Muslims, such as Adnan Oktar (also known as Harun Yahya), do not agree that one species can develop from another.

Since the 1980s, Turkey has been a site of strong advocacy for creationism, supported by American adherents.

There are several verses in the Qur'an which some modern writers have interpreted as being compatible with the expansion of the universe, Big Bang and Big Crunch theories:

> "Do not the Unbelievers see that the heavens and the earth were joined together (as one unit of creation), before we clove them asunder? We made from water every living thing. Will they not then believe?"

> "Moreover He comprehended in His design the sky, and it had been (as) smoke: He said to it and to the earth: 'Come ye together, willingly or unwillingly.' They said: 'We do come (together), in willing obedience.'"

> "With power and skill did We construct the Firmament: for it is We Who create the vastness of space."

> "The Day that We roll up the heavens like a scroll rolled up for books (completed),- even as We produced the first creation, so shall We produce a new one: a promise We have undertaken: truly shall We fulfil it."

Ahmadiyya

The Ahmadiyya movement actively promotes evolutionary theory. Ahmadis interpret scripture from the Qur'an to support the concept of macroevolution and give precedence to scientific theories. Furthermore, unlike orthodox Muslims, Ahmadis believe that mankind has gradually evolved from different species. Ahmadis regard Adam as being the first Prophet of God – as opposed to him being the first man on Earth. Rather than wholly adopting the theory of natural selection, Ahmadis promote the idea of a "guided evolution," viewing each stage of the evolutionary process as having been selectively woven by God. Mirza Tahir Ahmad, Fourth Caliph of the Ahmadiyya Muslim Community has stated in his magnum opus *Revelation, Rationality, Knowledge & Truth* (1998) that evolution did occur but only through God being the One who brings it about. It does not occur itself, according to the Ahmadiyya Muslim Community.

Judaism

For Orthodox Jews who seek to reconcile discrepancies between science and the creation myths in the Bible, the notion that science and the Bible should even be reconciled through traditional scientific means is questioned. To these groups, science is as true as the Torah and if there seems to be a problem, epistemological limits are to blame for apparently irreconcilable points. They point to discrepancies between what is expected and what actually is to demonstrate that things are not always as they appear. They note that even the root word for "world" in the Hebrew language means hidden. Just as they know from the Torah that God created man and trees and the light on its way from the stars in their observed state, so too can they know that the world was created in its over the six days of Creation that reflects progression to its currently-observed state, with the understanding that physical ways to verify this may eventually be identified. This knowledge has been advanced by Rabbi Dovid Gottlieb, former philosophy professor at Johns Hopkins University. Also, relatively old Kabbalistic sources from well before the scientifically apparent age of the universe was first determined are in close concord with modern scientific estimates of the age of the universe, according to Rabbi Aryeh Kaplan, and based on Sefer Temunah, an early kabbalistic work attributed to the first-century Tanna Nehunya ben HaKanah. Many kabbalists accepted the teachings of the Sefer HaTemunah, including the medieval Jewish scholar Nahmanides, his close student Isaac ben Samuel of Acre, and David ben Solomon ibn Abi Zimra. Other parallels are derived, among other sources, from Nahmanides, who expounds that there was a Neanderthal-like species with which Adam mated (he did this long before Neanderthals had even been discovered scientifically). Reform Judaism does not take the Torah as a literal text, but rather as a symbolic or open-ended work.

Some contemporary writers such as Rabbi Gedalyah Nadel have sought to reconcile the discrepancy between the account in the Torah, and scientific findings by arguing that each day referred to in the Bible was not 24 hours, but billions of years long. Others claim that the Earth was created a few thousand years ago, but was deliberately made to look as if it was five billion years old, e.g. by being created with ready made fossils. The best known exponent of this approach being Rabbi Menachem Mendel Schneerson Others state that although the world was physically created in six 24 hour days, the Torah accounts can be interpreted to mean that there was a period of billions of years before the six days of creation.

Bahá'í Faith

In the creation myth taught by Bahá'u'lláh, the Bahá'í Faith founder, the universe has "neither beginning nor ending," and that the component elements of the material world have always existed and will always exist. With regard to evolution and the origin of human beings, `Abdu'l-Bahá gave extensive comments on the subject when he addressed western audiences in the beginning of the 20th century. Transcripts of these comments can be found in *Some Answered Questions*, *Paris Talks* and *The Promulgation of Universal Peace*. `Abdu'l-Bahá described the human species as having evolved from a primitive form to modern man, but that the capacity to form human intelligence was always in existence.

Prevalence

Most vocal literalist creationists are from the US, and strict creationist views are much less common in other developed countries. According to a study published in *Science*, a survey of the US,

Turkey, Japan and Europe showed that public acceptance of evolution is most prevalent in Iceland, Denmark and Sweden at 80% of the population. There seems to be no significant correlation between believing in evolution and understanding evolutionary science.

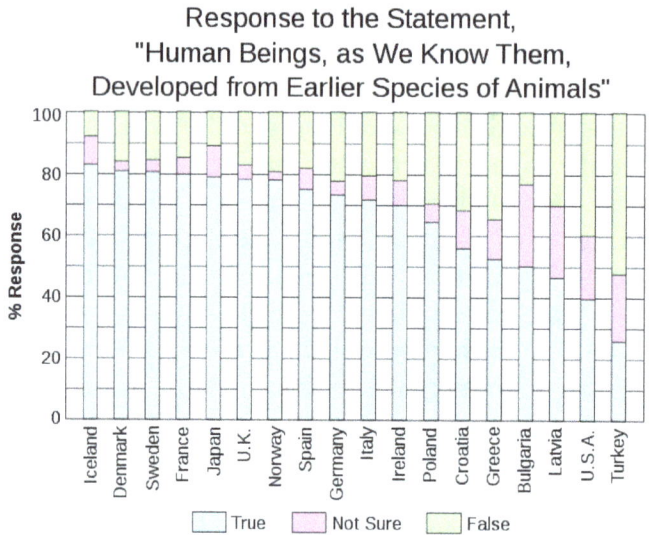

Views on human evolution in various countries

Australia

A 2009 Nielsen poll, showed that almost a quarter of Australians believe "the biblical account of human origins." Forty-two percent believe in a "wholly scientific" explanation for the origins of life, while 32 percent believe in an evolutionary process "guided by God."

Canada

A 2012 survey, by Angus Reid Public Opinion revealed that 61 percent of Canadians believe in evolution. The poll asked "Where did human beings come from – did we start as singular cells millions of year ago and evolve into our present form, or did God create us in his image 10,000 years ago?"

Europe

In Europe, literalist creationism is more widely rejected, though regular opinion polls are not available. Most people accept that evolution is the most widely accepted scientific theory as taught in most schools. In countries with a Roman Catholic majority, papal acceptance of evolutionary creationism as worthy of study has essentially ended debate on the matter for many people.

In the UK, a 2006 poll on the "origin and development of life", asked participants to choose between three different perspectives on the origin of life: 22% chose creationism, 17% opted for intelligent design, 48% selected evolutionary theory, and the rest did not know. A subsequent 2010 YouGov poll on the correct explanation for the origin of humans found that 9% opted for creationism, 12% intelligent design, 65% evolutionary theory and 13% didn't know. The former Archbishop of Canterbury Rowan Williams, head of the worldwide Anglican Communion, views the idea of teaching creationism in schools as a mistake.

In Italy, Education Minister Letizia Moratti wanted to retire evolution from the secondary school level; after one week of massive protests, she reversed her opinion.

There continues to be scattered and possibly mounting efforts on the part of religious groups throughout Europe to introduce creationism into public education. In response, the Parliamentary Assembly of the Council of Europe has released a draft report titled *The dangers of creationism in education* on June 8, 2007, reinforced by a further proposal of banning it in schools dated October 4, 2007.

Serbia suspended the teaching of evolution for one week in September 2004, under education minister Ljiljana Čolić, only allowing schools to reintroduce evolution into the curriculum if they also taught creationism. "After a deluge of protest from scientists, teachers and opposition parties" says the BBC report, Čolić's deputy made the statement, "I have come here to confirm Charles Darwin is still alive" and announced that the decision was reversed. Čolić resigned after the government said that she had caused "problems that had started to reflect on the work of the entire government."

Poland saw a major controversy over creationism in 2006, when the Deputy Education Minister, Mirosław Orzechowski, denounced evolution as "one of many lies" taught in Polish schools. His superior, Minister of Education Roman Giertych, has stated that the theory of evolution would continue to be taught in Polish schools, "as long as most scientists in our country say that it is the right theory." Giertych's father, Member of the European Parliament Maciej Giertych, has opposed the teaching of evolution and has claimed that dinosaurs and humans co-existed.

United States

Anti-evolution car in Athens, Georgia

A 2017 poll by Pew Research found that 62% of Americans believe humans have evolved over time and 34% of Americans believe humans and other living things have existed in their present form since the beginning of time.

According to a 2014 Gallup poll, about 42% of Americans believe that "God created human beings pretty much in their present form at one time within the last 10,000 years or so." Another 31% believe that "human beings have developed over millions of years from less advanced forms of life, but God guided this process,"and 19% believe that "human beings have developed over millions of years from less advanced forms of life, but God had no part in this process."

Belief in creationism is inversely correlated to education; of those with postgraduate degrees, 74% accept evolution. In 1987, *Newsweek* reported: "By one count there are some 700 scientists with respectable academic credentials (out of a total of 480,000 U.S. earth and life scientists) who give credence to creation-science, the general theory that complex life forms did not evolve but appeared 'abruptly.'"

A 2000 poll for People for the American Way found 70% of the US public felt that evolution was compatible with a belief in God.

According to a study published in *Science*, between 1985 and 2005 the number of adult North Americans who accept evolution declined from 45% to 40%, the number of adults who reject evolution declined from 48% to 39% and the number of people who were unsure increased from 7% to 21%. Besides the US the study also compared data from 32 European countries, Turkey, and Japan. The only country where acceptance of evolution was lower than in the US was Turkey (25%).

According to a 2011 Fox News poll, 45% of Americans believe in Creationism, down from 50% in a similar poll in 1999. 21% believe in 'the theory of evolution as outlined by Darwin and other scientists' (up from 15% in 1999), and 27% answered that both are true (up from 26% in 1999).

In September 2012, educator and television personality Bill Nye spoke with the Associated Press and aired his fears about acceptance of creationism, believing that teaching children that creationism is the only true answer and without letting them understand the way science works will prevent any future innovation in the world of science. In February 2014, Nye defended evolution in the classroom in a debate with creationist Ken Ham on the topic of whether creation is a viable model of origins in today's modern, scientific era.

Education Controversies

The Truth fish, one of the many creationist responses to the Darwin fish

In the US, creationism has become centered in the political controversy over creation and evolution in public education, and whether teaching creationism in science classes conflicts with the separation of church and state. Currently, the controversy comes in the form of whether advocates of the intelligent design movement who wish to "Teach the Controversy" in science classes have conflated science with religion.

People for the American Way polled 1500 North Americans about the teaching of evolution and creationism in November and December 1999. They found that most North Americans were not familiar with Creationism, and most North Americans had heard of evolution, but many did not fully understand the basics of the theory. The main findings were:

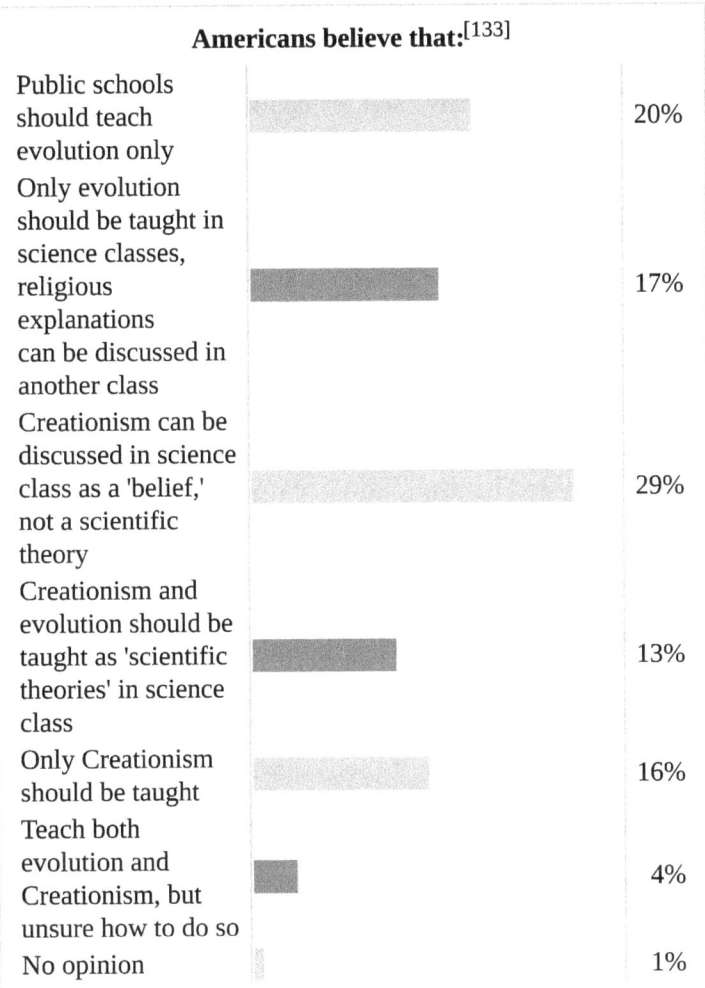

Americans believe that:[133]

Public schools should teach evolution only	20%
Only evolution should be taught in science classes, religious explanations can be discussed in another class	17%
Creationism can be discussed in science class as a 'belief,' not a scientific theory	29%
Creationism and evolution should be taught as 'scientific theories' in science class	13%
Only Creationism should be taught	16%
Teach both evolution and Creationism, but unsure how to do so	4%
No opinion	1%

In such political contexts, creationists argue that their particular religiously based origin belief is superior to those of other belief systems, in particular those made through secular or scientific rationale. Political creationists are opposed by many individuals and organizations who have made detailed critiques and given testimony in various court cases that the alternatives to scientific reasoning offered by creationists are opposed by the consensus of the scientific community.

Criticism

Christian Criticism

Most Christians disagree with the teaching of creationism as an alternative to evolution in schools. Several religious organizations, among them the Catholic Church, hold that their faith does not conflict with the scientific consensus regarding evolution. The Clergy Letter Project, which has collected more than 13,000 signatures, is an "endeavor designed to demonstrate that religion and science can be compatible."

In his 2002 article "Intelligent Design as a Theological Problem," George Murphy argues against the view that life on Earth, in all its forms, is direct evidence of God's act of creation (Murphy quotes Phillip E. Johnson's claim that he is speaking "of a God who acted openly and left his fingerprints on all the evidence."). Murphy argues that this view of God is incompatible with the Christian understanding of God as "the one revealed in the cross and resurrection of Christ." The basis of this theology is Isaiah 45:15, "Verily thou art a God that hidest thyself, O God of Israel, the Saviour."

Murphy observes that the execution of a Jewish carpenter by Roman authorities is in and of itself an ordinary event and did not require divine action. On the contrary, for the crucifixion to occur, God had to limit or "empty" Himself. It was for this reason that Paul the Apostle wrote, in Philippians 2:5-8:

> "Let this mind be in you, which was also in Christ Jesus: Who, being in the form of God, thought it not robbery to be equal with God: But made himself of no reputation, and took upon him the form of a servant, and was made in the likeness of men: And being found in fashion as a man, he humbled himself, and became obedient unto death, even the death of the cross."

Murphy concludes that,

"Just as the Son of God limited himself by taking human form and dying on a cross, God limits divine action in the world to be in accord with rational laws which God has chosen. This enables us to understand the world on its own terms, but it also means that natural processes hide God from scientific observation."

For Murphy, a theology of the cross requires that Christians accept a *methodological* naturalism, meaning that one cannot invoke God to explain natural phenomena, while recognizing that such acceptance does not require one to accept a *metaphysical* naturalism, which proposes that nature is all that there is.

The Jesuit priest George Coyne has stated that is "unfortunate that, especially here in America, creationism has come to mean...some literal interpretation of Genesis." He argues that "...Judaic-Christian faith is radically creationist, but in a totally different sense. It is rooted in belief that everything depends on God, or better, all is a gift from God."

Teaching of Creationism

Other Christians have expressed qualms about teaching creationism. In March 2006, then Archbishop of Canterbury Rowan Williams, the leader of the world's Anglicans, stated his discomfort about teaching creationism, saying that creationism was "a kind of category mistake, as if the Bible were a theory like other theories." He also said: "My worry is creationism can end up reducing the doctrine of creation rather than enhancing it." The views of the Episcopal Church – a major American-based branch of the Anglican Communion – on teaching creationism resemble those of Williams.

The National Science Teachers Association is opposed to teaching creationism as a science, as is the Association for Science Teacher Education, the National Association of Biology Teachers, the American Anthropological Association, the American Geosciences Institute, the Geological Society of America, the American Geophysical Union, and numerous other professional teaching and scientific societies.

In April 2010, the American Academy of Religion issued *Guidelines for Teaching About Religion in K - 12 Public Schools in the United States* which included guidance that creation science or intelligent design should not be taught in science classes, as "Creation science and intelligent design represent worldviews that fall outside of the realm of science that is defined as (and limited to) a method of inquiry based on gathering observable and measurable evidence subject to specific principles of reasoning." However, they, as well as other "worldviews that focus on speculation regarding the origins of life represent another important and relevant form of human inquiry that is appropriately studied in literature or social sciences courses. Such study, however, must include a diversity of worldviews representing a variety of religious and philosophical perspectives and must avoid privileging one view as more legitimate than others."

Randy Moore and Sehoya Cotner, from the biology program at the University of Minnesota, reflect on the relevance of teaching creationism in the article *The Creationist Down the Hall: Does It Matter When Teachers Teach Creationism?* They conclude that "Despite decades of science education reform, numerous legal decisions declaring the teaching of creationism in public-school science classes to be unconstitutional, overwhelming evidence supporting evolution, and the many denunciations of creationism as nonscientific by professional scientific societies, creationism remains popular throughout the United States."

Scientific Criticism

Science is a system of knowledge based on observation, empirical evidence, and the development of theories that yield testable explanations and predictions of natural phenomena. By contrast, creationism is often based on literal interpretations of the narratives of particular religious texts. Some creationist beliefs involve purported forces that lie outside of nature, such as supernatural intervention, and often do not allow predictions at all. Therefore, these can neither be confirmed nor disproved by scientists. However, many creationist beliefs can be framed as testable predictions about phenomena such as the age of the Earth, its geological history and the origins, distributions and relationships of living organisms found on it. Early science incorporated elements of these beliefs, but as science developed these beliefs were gradually falsified and were replaced with understandings based on accumulated and reproducible evidence that often allows the accurate prediction of future results.

Some scientists, such as Stephen Jay Gould, consider science and religion to be two compatible and complementary fields, with authorities in distinct areas of human experience, so-called non-overlapping magisteria. This view is also held by many theologians, who believe that ultimate origins and meaning are addressed by religion, but favor verifiable scientific explanations of natural phenomena over those of creationist beliefs. Other scientists, such as Richard Dawkins, reject the non-overlapping magisteria and argue that, in disproving literal interpretations of creationists, the scientific method also undermines religious texts as a source of truth. Irrespective of this diversity in viewpoints, since creationist beliefs are not supported by empirical evidence, the scientific consensus is that any attempt to teach creationism as science should be rejected.

Organizations

Creationism (in general)

- American Scientific Affiliation

- Christians in Science

Young Earth Creationism

- Answers in Genesis, a group promoting young Earth creationism

- Creation Ministries International, an organisation promoting biblical creation

- Creation Research Society

- Institute for Creation Research

- The Way of the Master

Old Earth Creationism

- Old Earth Ministries (OEM), formerly Answers In Creation (AIC), led by Greg Neyman

- Reasons to Believe, led by Hugh Ross

Intelligent design

- Access Research Network

- Centre for Intelligent Design

- Center for Science and Culture

Evolutionary creationism

- BioLogos Foundation

Evolution

- National Center for Science Education

- TalkOrigins Archive

- The Panda's Thumb (blog)

- ScienceBlogs

- Why Evolution is True (Jerry Coyne's website)

Creation Myth

A creation myth (also called a cosmogonic myth) is a symbolic narrative of how the world began and how people first came to inhabit it. While in popular usage the term *myth* often refers to false or fanciful stories, formally, it does not imply falsehood. Cultures generally regard their creation myths as true. In the society in which it is told, a creation myth is usually regarded as conveying profound truths, metaphorically, symbolically and sometimes in a historical or literal sense. They are commonly, although not always, considered cosmogonical myths – that is, they describe the ordering of the cosmos from a state of chaos or amorphousness.

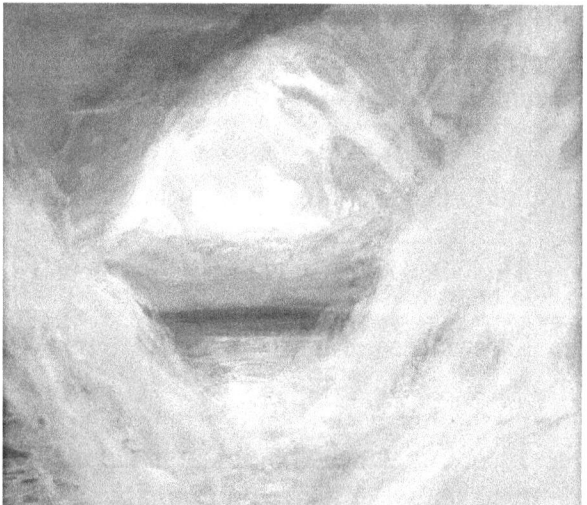

The Creation (c. 1896–1902) by James Tissot

Creation myths often share a number of features. They often are considered sacred accounts and can be found in nearly all known religious traditions. They are all stories with a plot and characters who are either deities, human-like figures, or animals, who often speak and transform easily. They are often set in a dim and nonspecific past that historian of religion Mircea Eliade termed *in illo tempore* ("at that time"). Creation myths address questions deeply meaningful to the society that shares them, revealing their central worldview and the framework for the self-identity of the culture and individual in a universal context.

Creation myths develop in oral traditions and therefore typically have multiple versions; found throughout human culture, they are the most common form of myth.

Definitions

Creation myth definitions from modern references:

- A "symbolic narrative of the beginning of the world as understood in a particular tradition and community. Creation myths are of central importance for the valuation of the world, for the orientation of humans in the universe, and for the basic patterns of life and culture."

- "Creation myths tell us how things began. All cultures have creation myths; they are our primary myths, the first stage in what might be called the psychic life of the species. As cultures, we identify ourselves through the collective dreams we call creation myths, or cosmogonies. ... Creation myths explain in metaphorical terms our sense of who we are in the context of the world, and in so doing they reveal our real priorities, as well as our real prejudices. Our images of creation say a great deal about who we are."

- A "philosophical and theological elaboration of the primal myth of creation within a religious community. The term myth here refers to the imaginative expression in narrative form of what is experienced or apprehended as basic reality ... The term creation refers to the beginning of things, whether by the will and act of a transcendent being, by emanation from some ultimate source, or in any other way."

Religion professor Mircea Eliade defined the word *myth* in terms of creation:

Myth narrates a sacred history; it relates an event that took place in primordial Time, the fabled time of the "beginnings." In other words, myth tells how, through the deeds of Supernatural Beings, a reality came into existence, be it the whole of reality, the Cosmos, or only a fragment of reality – an island, a species of plant, a particular kind of human behavior, an institution.

Meaning and Function

All creation myths are in one sense etiological because they attempt to explain how the world was formed and where humanity came from.

Ethnologists and anthropologists who study these myths say that in the modern context theologians try to discern humanity's meaning from revealed truths and scientists investigate cosmology with the tools of empiricism and rationality, but creation myths define human reality in very different terms. In the past historians of religion and other students of myth thought of them as forms of primitive or early-stage science or religion and analyzed them in a literal or logical sense. However they are today seen as symbolic narratives which must be understood in terms of their own cultural context. Charles Long writes, "The beings referred to in the myth – gods, animals, plants – are forms of power grasped existentially. The myths should not be understood as attempts to work out a rational explanation of deity."

While creation myths are not literal explications they do serve to define an orientation of humanity in the world in terms of a birth story. They are the basis of a worldview that reaffirms and guides how people relate to the natural world, to any assumed spiritual world, and to each other. The creation myth acts as a cornerstone for distinguishing primary reality from relative reality, the origin and nature of being from non-being. In this sense they serve as a philosophy of life but one expressed and conveyed through symbol rather than systematic reason. And in this sense they go beyond etiological myths which mean to explain specific features in religious rites, natural phenomena or cultural life. Creation myths also help to orient human beings in the world, giving them a sense of their place in the world and the regard that they must have for humans and nature.

Historian David Christian has summarised issues common to multiple creation myths:

Each beginning seems to presuppose an earlier beginning. ... Instead of meeting a single starting point, we encounter an infinity of them, each of which poses the same problem. ... There are no entirely satisfactory solutions to this dilemma. What we have to find is not a solution but some way of dealing with the mystery .... And we have to do so using words. The words we reach for, from *God* to *gravity*, are inadequate to the task. So we have to use language poetically or symbolically; and such language, whether used by a scientist, a poet, or a shaman, can easily be misunderstood.

Classification

Mythologists have applied various schemes to classify creation myths found throughout human cultures. Eliade and his colleague Charles Long developed a classification based on some common motifs that reappear in stories the world over. The classification identifies five basic types:

In Maya religion, the dwarf was an embodiment of the Maize God's helpers at creation.

Brahmā, the Hindu *deva* of creation, emerges from a lotus risen from
the navel of Viṣṇu, who lies with Lakshmi on the serpent Ananta Shesha.

- Creation *ex nihilo* in which the creation is through the thought, word, dream or bodily secretions of a divine being.

- Earth diver creation in which a diver, usually a bird or amphibian sent by a creator, plunges to the seabed through a primordial ocean to bring up sand or mud which develops into a terrestrial world.

- Emergence myths in which progenitors pass through a series of worlds and metamorphoses until reaching the present world.

- Creation by the dismemberment of a primordial being.

- Creation by the splitting or ordering of a primordial unity such as the cracking of a cosmic egg or a bringing order from chaos.

Marta Weigle further developed and refined this typology to highlight nine themes, adding elements such as *deus faber*, a creation crafted by a deity, creation from the work of two creators working together or against each other, creation from sacrifice and creation from division/conjugation, accretion/conjunction, or secretion.

An alternative system based on six recurring narrative themes was designed by Raymond Van Over:

- Primeval abyss, an infinite expanse of waters or space.

- Originator deity which is awakened or an eternal entity within the abyss.

- Originator deity poised above the abyss.

- Cosmic egg or embryo.

- Originator deity creating life through sound or word.

- Life generating from the corpse or dismembered parts of an originator deity.

Ex Nihilo

Creation on the exterior shutters of Hieronymus Bosch's triptych *The Garden of Earthly Delights* (1480–90)

Creation *ex nihilo* (Latin for "out of nothing"), also known as "creation *de novo* (Latin for "from the new")", is a common type of mythical creation. *Ex nihilo* creation is found in creation stories from ancient Egypt, the Rig Veda, the Bible and the Quran, and many animistic cultures in Africa, Asia, Oceania and North America. The Debate between sheep and grain is an example of an even earlier form of *ex nihilo* creation myth from ancient Sumer. In most of these stories the world is brought into being by the speech, dream, breath, or pure thought of a creator but creation ex nihilo may also take place through a creator's bodily secretions.

The literal translation of the phrase *ex nihilo* is "from nothing" but in many creation myths the line is blurred whether the creative act would be better classified as a creation *ex nihilo* or creation from chaos. In *ex nihilo* creation myths the potential and the substance of creation springs from within the creator. Such a creator may or may not be existing in physical surroundings such as darkness or water, but does not create the world from them, whereas in creation from chaos the substance used for creation is pre-existing within the unformed void.

A well known example of an ex nihilo creation myth is the one found in the Bible.

Creation from Chaos

In creation from chaos myth, initially there is nothing but a formless, shapeless expanse. In these stories the word "chaos" means "disorder", and this formless expanse, which is also sometimes called a void or an abyss, contains the material with which the created world will be made. Chaos may be described as having the consistency of vapor or water, dimensionless, and sometimes salty or muddy. These myths associate chaos with evil and oblivion, in contrast to "order" (*cosmos*) which is the good. The act of creation is the bringing of order from disorder, and in many of these cultures it is believed that at some point the forces preserving order and form will weaken and the world will once again be engulfed into the abyss.

World Parent

In one Maori creation myth, the primal couple are Rangi and Papa, depicted holding each other in a tight embrace.

There are two types of world parent myths, both describing a separation or splitting of a primeval entity, the world parent or parents. One form describes the primeval state as an eternal union of two parents, and the creation takes place when the two are pulled apart. The two parents are commonly identified as Sky (usually male) and Earth (usually female) who in the primeval state were so tightly bound to each other that no offspring could emerge. These myths often depict creation as the result of a sexual union, and serve as genealogical record of the deities born from it.

In the second form of world parent myth, creation itself springs from dismembered parts of the body of the primeval being. Often in these stories the limbs, hair, blood, bones or organs of the primeval being are somehow severed or sacrificed to transform into sky, earth, animal or plant life, and other worldly features. These myths tend to emphasize creative forces as animistic in nature rather than sexual, and depict the sacred as the elemental and integral component of the natural world. One example of this is the Norse creation myth described in Völuspá.

Emergence

In the kiva of both ancient and present-day Pueblo peoples, the sipapu is a small round hole in the floor that represents the portal through which the ancestors first emerged. (The larger hole is a fire pit, here in a ruin from the Mesa Verde National Park.)

In emergence myths humanity emerges from another world into the one they currently inhabit. The previous world is often considered the womb of the earth mother, and the process of emergence is likened to the act of giving birth. The role of midwife is usually played by a female deity, like the spider woman of Native American mythology. Male characters rarely figure into these stories, and scholars often consider them in counterpoint to male-oriented creation myths, like those of the *ex nihilo* variety.

Emergence myths commonly describe the creation of people and/or supernatural beings as a staged ascent or metamorphosis from nascent forms through a series of subterranean worlds to arrive at their current place and form. Often the passage from one world or stage to the next is impelled by inner forces, a process of germination or gestation from earlier, embryonic forms. The genre is most commonly found in Native American cultures where the myths frequently link the final emergence of people from a hole opening to the underworld to stories about their subsequent migrations and eventual settlement in their current homelands.

Earth-diver

The earth-diver is a common character in various traditional creation myths. In these stories a supreme being usually sends an animal into the primal waters to find bits of sand or mud with which to build habitable land. Some scholars interpret these myths psychologically while others interpret them cosmogonically. In both cases emphasis is placed on beginnings emanating from the depths. Earth-diver myths are common in Native American folklore but can be found among the Chukchi and Yukaghir, the Tatars and many Finno-Ugrian traditions. The pattern of distribution of these stories suggest they have a common origin in the eastern Asiatic coastal region, spreading as peoples migrated west into Siberia and east to the North American continent.

Characteristic of many Native American myths, earth-diver creation stories begin as beings and potential forms linger asleep or suspended in the primordial realm. The earth-diver is among the first of them to awaken and lay the necessary groundwork by building suitable lands where the coming creation will be able to live. In many cases, these stories will describe a series of failed attempts to make land before the solution is found.

References

- Gravrand, Henry, "La civilisation sereer : Pangool", vol. 2, Les Nouvelles Editions Africaines du Senegal, (1990) pp 20-21, 149-155, ISBN 2-7236-1055-1

- Tan, Piya. "Aggañña Sutta" (PDF). The Dharmafarers. Retrieved 7 May 2015. Then the female developed female organs,87 and the male developed male organs

- Aviezer, Nathan (1990). In the Beginning—: Biblical Creation and Science. Hoboken, NJ: KTAV Publishing House. ISBN 0-88125-328-6. LCCN 89049127. OCLC 20800545

- Barlow, Nora, ed. (1963). "Darwin's Ornithological Notes". Bulletin of the British Museum (Natural History) Historical Series. London: Trustees of the British Museum. 2 (7): 201–278. ISSN 0068-2306. Retrieved 2009-06-10

- Dawkins, Richard (2006). The God Delusion. London: Bantam Press. ISBN 978-0-5930-5548-9. LCCN 2006015506. OCLC 70671839

- Bucaille, Maurice (1976). The Qur'an and Modern Science (Booklet). Riyadh, Kingdom of Saudi Arabia: Co-operative Offices for Call & Guidance at Al-Badiah & Industrial Area. OCLC 52246825. Retrieved 2014-03-21

- Draper, Paul R. (2005). "God, Science, and Naturalism". In Wainwright, William J. The Oxford Handbook of Philosophy of Religion. Oxford; New York: Oxford University Press. ISBN 978-0-1951-3809-2. LCCN 2004043890. OCLC 54542845. doi:10.1093/0195138090.003.0012

- Masood, Steven (1994) [Originally published 1986]. Jesus and the Indian Messiah. Oldham, England: Word of Life. ISBN 1-898868-00-X. LCCN 94229476. OCLC 491161526

- "Worldwide Adherents of All Religions by Six Continental Areas, Mid-2002". Encyclopædia Britannica. 2002. Archived from the original on 2006-02-21. Retrieved 31 May 2006

- Christian, David (2004). Maps of Time: An Introduction to Big History. California World History Library. 2. University of California Press. pp. 17–18. ISBN 9780520931923

Permissions

We would like to thank the editorial team for lending their expertise to make the book truly unique. They have played a crucial role in the development of this book. Without their invaluable contributions this book wouldn't have been possible. They have made vital efforts to compile up to date information on the varied aspects of this subject to make this book a valuable addition to the collection of many professionals and students.

This book was conceptualized with the vision of imparting up-to-date and integrated information in this field. To ensure the same, a matchless editorial board was set up. Every individual on the board went through rigorous rounds of assessment to prove their worth. After which they invested a large part of their time researching and compiling the most relevant data for our readers.

The editorial board has been involved in producing this book since its inception. They have spent rigorous hours researching and exploring the diverse topics which have resulted in the successful publishing of this book. They have passed on their knowledge of decades through this book. To expedite this challenging task, the publisher supported the team at every step. A small team of assistant editors was also appointed to further simplify the editing procedure and attain best results for the readers.

Apart from the editorial board, the designing team has also invested a significant amount of their time in understanding the subject and creating the most relevant covers. They scrutinized every image to scout for the most suitable representation of the subject and create an appropriate cover for the book.

The publishing team has been an ardent support to the editorial, designing and production team. Their endless efforts to recruit the best for this project, has resulted in the accomplishment of this book. They are a veteran in the field of academics and their pool of knowledge is as vast as their experience in printing. Their expertise and guidance has proved useful at every step. Their uncompromising quality standards have made this book an exceptional effort. Their encouragement from time to time has been an inspiration for everyone.

The publisher and the editorial board hope that this book will prove to be a valuable piece of knowledge for students, practitioners and scholars across the globe.

Index

www.ingramcontent.com/pod-product-compliance
Lightning Source LLC
Chambersburg PA
CBHW080408190526
45161CB00003B/172